Essential Maths Skills
for AS/A-level
Economics

Peter Davis and Tracey Joad

PHILIP ALLAN FOR
HODDER
EDUCATION
AN HACHETTE UK COMPANY

Philip Allan, an imprint of Hodder Education, an Hachette UK company,
Blenheim Court, George Street, Banbury, Oxfordshire OX16 5BH

Orders

Bookpoint Ltd, 130 Park Drive, Milton Park, Abingdon, Oxfordshire OX14 4SE
tel: 01235 827827
fax: 01235 400401
e-mail: education@bookpoint.co.uk

Lines are open 9.00 a.m.–5.00 p.m., Monday to Saturday, with a 24-hour message
answering service. You can also order through the Hodder Education website:
www.hoddereducation.co.uk

© Peter Davis and Tracey Joad 2016

ISBN 978-1-4718-6350-9

First printed 2016
Impression number 5 4 3 2 1
Year 2020 2019 2018 2017 2016

Typeset by Aptara, Inc.

Cover illustration: Barking Dog Art

Printed in Spain

Hachette UK's policy is to use papers that are natural, renewable and recyclable products and
made from wood grown in sustainable forests. The logging and manufacturing processes are
expected to conform to the environmental regulations of the country of origin.

Contents

The listed content is assessed by the awarding bodies AQA, OCR, Pearson Edexcel, WJEC/Eduqas and CCEA at AS and A-level.

Introduction

Economics is an exciting subject to study, and most students enjoy the many different skills they develop throughout the course. Because economics is a social science, economists must carefully analyse and evaluate how decisions are made by consumers, firms and governments as well as the effects these decisions may have on the economy as a whole. Mathematics is an essential tool that real-world economists use to help them do this.

The AS and A-level economics courses now make clear the range of quantitative (i.e. mathematical) skills that you need to be competent in. These skills will be used throughout the course and across most topic areas. For the AS course, a minimum of 15% of the total marks available will be allocated to these mathematical skills, and at A-level it is 20% of the overall marks.

The main message is: **don't panic — just be prepared!** This book will start at the beginning: it assumes little previous knowledge and will carefully build up your mathematical skills. Each unit links to a particular quantitative skill listed in the specifications, so by working carefully through the book you will be able to answer confidently the range of 'mathematical' questions which are likely to come up in your economics exams. Even if you are already a good mathematician, this book will reinforce the necessary skills and provide you with plenty of examples of how each skill might be applied in a relevant economic context. In each unit of the book you will find:

- an introductory explanation with worked examples
- guided questions which take you through the calculations step by step to help you build up your knowledge and confidence
- a variety of practice questions

The exam-style questions at the end of the book will give you some idea of what to expect in the exams.

Full worked solutions to the guided and practice questions and exam-style questions can be found online at www.hoddereducation.co.uk/essentialmathsanswers.

1 Fractions and ratios

Fractions

A fraction is a part of a whole. It is usually written as one number over another number, separated by a horizontal line.

- The bottom number (the denominator) shows how many equal parts a whole has been divided into.
  For example, if a pizza is divided into six equal slices, then the denominator is 6.
- The top number (the numerator) says how many of those parts are being considered.
  For example, if you eat three slices of the above pizza, then the numerator is 3 and you have eaten $\frac{3}{6}$ of the pizza.

Fractions are often used in calculations to find a portion of a total amount. For example, calculating $\frac{2}{5}$ of £240 means splitting £240 into five equal parts and then taking two of these parts.

Simplifying fractions

Figure 1.1

The total number of mobile phones in the container is 6. There are 2 red phones and 4 black phones. This means that $\frac{2}{6}$ of the phones are red and $\frac{4}{6}$ are black. Both of these fractions can be simplified.

Figure 1.2 can help you simplify fractions. Equivalent fractions are shown on the coloured bricks. The bricks in each row add up to the whole.

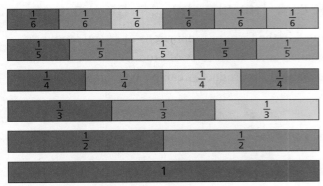

Figure 1.2 Equivalent fractions

From the diagram you can see that:

- $\frac{2}{4}$ is equivalent to $\frac{1}{2}$ (a half)
- $\frac{2}{6}$ is equivalent to $\frac{1}{3}$ (one-third)
- $\frac{4}{6}$ is equivalent to $\frac{2}{3}$ (two-thirds)

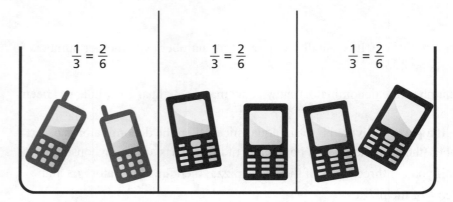

Figure 1.3

So $\frac{1}{3}$ (one-third) of the phones in the container are red and $\frac{2}{3}$ (two-thirds) are black. This can also be expressed using the word 'proportion': the proportion of red phones in the container is $\frac{1}{3}$.

To simplify fractions without using 'bricks' as shown in Figure 1.2, think if there is a number that divides into **both** the numerator and the denominator (a 'common factor'). For example, with the fraction $\frac{4}{6}$, the number 2 divides into both top and bottom: $4 \div 2 = 2$ and $6 \div 2 = 3$, so the fraction becomes $\frac{2}{3}$.

Converting fractions to decimals

Sometimes it may be more convenient to write a fraction as a decimal. To do this, simply divide the numerator (top number) by the denominator (bottom number).

For example, $\frac{2}{6}$ means 2 divided by 6, so

$$\frac{2}{6} = 2 \div 6 = 0.3333 \ldots$$

Note that this is a recurring decimal, i.e. after the decimal point there is a repeating pattern of digits that never terminates.

We can write the result as 0.33 (2 d.p.), which means '0.33 to two decimal places'.

Using fractions to work out 'how many' or 'how much'

In economics, information is often given in the form of proportions of a total amount, for example, 'a fifth of total sales revenue comes from sales of product A'. So, if the total sales revenue is £400 000, then the revenue from product A can be calculated as

$$\frac{1}{5} \times £400\,000$$

(a fifth of £400 000 means $\frac{1}{5}$ times £400 000)

On a calculator this could be done as 1 divided by 5 and then the answer multiplied by 400 000. This gives an answer of £80 000.

Understanding and being confident in calculating with fractions will help you greatly throughout your AS/A-level economics course. Situations in which fractions are frequently used include:

- assessing the significance of some factor by considering what proportion of the total it accounts for, e.g. what proportion of the workforce is from a particular age group
- calculating percentages (a percentage is just a fraction where the denominator is 100 — see Unit 2)
- elasticity calculations (see Unit 2)

Ratios

Fractions and ratios are closely related, but instead of comparing a 'part' with a 'whole' as fractions do, ratios compare the relative amounts of two or more items.

In the example shown in Figure 1.1, there are 2 red phones and 4 black phones in the container. So the ratio of red phones to black phones is $2:4$. The colon (:) means 'compared to'.

Simplifying ratios

Ratios can be simplified in the same way that fractions can be simplified: divide the numbers on both sides of the colon by the same number. It is often useful to simplify so that one of the sides becomes 1.

For example, if you divide both sides of the ratio $2:4$ by 2, you get the simplified ratio $1:2$.

Figure 1.4

The original ratio $2:4$ means that for every 2 red phones there are 4 black ones. This is the same as saying 'for every 1 red phone there are 2 black ones'.

You could also divide both sides of the ratio $2:4$ by 4. This gives

$$(2 \div 4):(4 \div 4) = \frac{1}{2}:1$$

Here are some examples of how ratios can arise in your economics course:

- Ratios are particularly useful when you need to make a quick comparison of two quantities, i.e. how much bigger or smaller one amount is compared to another. For example:
  - comparing living standards of two countries by comparing their GDP per head
  - comparing price inflation with wage inflation to assess cost-of-living changes over time
  - comparing average income of the richest 10% and poorest 10% of a country's population

- At A-level, opportunity cost ratios are used to understand the law of comparative advantage, which is an international trade topic.

(A) Worked examples

a **A firm pays £200 per week on rent for its premises. It spends £800 per week on all its other costs.**

  i **What proportion of the firm's total costs is rent?**

  Step 1: to calculate a proportion (fraction), first find the amount of the 'whole'.
  Total costs = £200 + £800 = £1000

  Step 2: use the amount found in Step 1 as the denominator and the 'part' you are interested in as the numerator to set up the fraction.

  Proportion spent on rent is

  $$\frac{\text{rent cost}}{\text{total costs}} = \frac{200}{1000}$$

  Step 3: simplify the fraction if possible.
  The number 200 can be divided into both top and bottom of the fraction: $200 \div 200 = 1$ and $1000 \div 200 = 5$. Therefore, the proportion spent on rent is $\frac{1}{5}$.

  ii **What is the ratio of rent to other costs?**

  Step 1: set up the ratio by writing down the relevant amounts separated by a colon.

  Rent : Other costs
  = 200 : 800

  Step 2: simplify the ratio if possible.

  Dividing both sides by 200 gives

  $(200 \div 200) : (800 \div 200) = 1 : 4$

  This means that for every £1 spent on rent, £4 is spent on other costs.

b **An economy has a population of 60 million. Two-thirds of the population are of working age. (The working-age population is defined by the World Bank as those individuals from 15 to 64 years old. Dependents are defined as individuals younger than 15 or older than 64.)**

  i **Calculate the size of the working-age population.**

  Two-thirds of 60 million is calculated as
  $$\frac{2}{3} \times 60 \text{ million} = 40 \text{ million}$$

  On a calculator this can be done as 2 divided by 3 and then the answer multiplied by 60.

ii **Calculate the proportion of the population who are dependents and the total number of dependents.**

- If two-thirds of the population are of working age, this leaves one-third, i.e. $\frac{1}{3}$, of the population who are not of working age.

- Since individuals are either dependents or part of the working-age population, the proportion of the population who are dependents must be $\frac{1}{3}$.

The number of dependents is

$$\frac{1}{3} \times 60 \text{ million} = 20 \text{ million}$$

Alternatively, find this number from 60 million − 40 million = 20 million.

B Guided questions

Copy out the workings and complete the answers on a separate piece of paper.

1 **In 2013 the GDP per capita (GDP per head), in US dollars at purchasing power parity (PPP), was approximately $2400 for Tanzania and $38 400 for the UK.**

   a **What was the ratio of the UK's GDP per capita to Tanzania's GDP per capita in 2013?**

   Step 1: set up the ratio by writing down the relevant amounts separated by a colon.

   UK's GDP per capita : Tanzania's GDP per capita

   = _____ : _____

   Step 2: simplify the ratio.

   Divide both sides by Tanzania's GDP per capita.

   UK's GDP per capita : Tanzania's GDP per capita

   = _____ : _____

   b **What conclusions can be drawn from this information?**

   The simplified ratio shows that in 2013 the average income in the UK was approximately _____ times the average income in Tanzania.

   The data is given at purchasing power parity, so cost-of-living differences have been accounted for. This means that the standard of living in terms of the quantity of goods that an individual can buy was approximately _____ times higher in the UK than in Tanzania.

2 **With the same amount of resources, country A can produce 15 cars or 60 bicycles, whereas country B can produce 10 cars or 50 bicycles.**

   a **For each country, work out the opportunity cost of producing one car (in terms of the number of bicycles that could have been produced instead). Which country has the smaller opportunity cost of producing one car?**

- **Opportunity cost** is a concept which is covered early in the A-level economics course.
- It is the value of the next best alternative foregone, and is often used in decision making.

Step 1: set up ratios by writing down the relevant amounts separated by a colon.

In this case, you want to show, for each country, how many cars could be produced compared to how many bicycles could be produced.

	Cars	:	Bicycles
Country A	15	:	60
Country B	___	:	___

Step 2: simplify the ratios.

Because you need to find the opportunity cost of producing one car, divide both sides of each ratio by the number of cars to convert that quantity to 1. This will allow us to see what number of bicycles would need to be sacrificed if the resources are used to produce one car.

For country A, divide both sides of the ratio by 15.
For country B, divide both sides by ___

Then the ratios become

	Cars	:	Bicycles
Country A	1	:	___
Country B	1	:	___

Step 3: draw conclusions from the simplified ratios.

Country A would have to sacrifice ___ bicycles to produce one car, whereas country B would have to sacrifice ___ bicycles. Therefore country ___ has the smaller opportunity cost of producing one car.

b **Work out the opportunity cost of producing one bicycle for each country.**

To find the opportunity cost of producing one bicycle, divide both sides of each original ratio by the number of bicycles to convert that quantity to 1. This will allow us to see what number of cars would need to be sacrificed if the resources are used to produce one bicycle.

This time the simplified ratios are

	Cars	:	Bicycles
Country A	___	:	1
Country B	___	:	1

Country A would have to sacrifice ___ cars to produce one bicycle, whereas country B would have to sacrifice ___ cars. Therefore country ___ has the smaller opportunity cost of producing one bicycle.

 Full worked solutions at **www.hoddereducation.co.uk/essentialmathsanswers**

If you're not sure which of the fractions is smaller, convert them into decimals for easier comparison, or look at Figure 1.2.

C Practice questions

3   A household in the UK spends £480 per week. Some of the main categories of spending are listed below.

Table 1.1 Household spending per week

Category	Amount spent
Housing, fuel and power	£80
Transport	£60
Food and non-alcoholic drinks	£60
Recreation and culture	£60
Restaurants and hotels	£40
Household goods and services	£30

a   What proportion of total spending is spent on each category listed in Table 1.1?

b   Find the ratio of spending on restaurants and hotels compared to spending on recreation and culture. Present this ratio in a simplified form.

c   If this household spends $\frac{2}{5}$ of its transport spending on 'public transport', work out how many pounds it spends per week on public transport and how much on other forms of transport.

4   A manufacturer of mobile phone accessories has set up his equipment so he can produce three black phone cases and two grey phone cases per minute.

a   What proportion of the phone cases produced each minute are black?

b   Write down a ratio that shows the number of black phone cases compared to the number of grey phone cases produced per minute.

c   Simplify the ratio in part b to show the number of grey phone cases produced for every black phone case per minute.

d   Simplify the ratio in part b to show the number of black phone cases produced for every grey phone case per minute.

5   In 2013 the GDP per capita, in US dollars and PPP-adjusted, was approximately $25 500 for Greece and $53 550 for the USA.

Write down the ratio of GDP per capita for Greece compared to that for the USA.

Use this ratio to describe how much higher the standard of living in the USA seems to be.

6   With an identical amount of resources, country A can produce 30 tractors or 60 lorries, whereas country B can produce 40 tractors or 50 lorries.

a   For each country, work out the opportunity cost of producing one tractor. (To do this, set out the ratios and then simplify them in an appropriate way.)

b   For each country, work out the opportunity cost of producing one lorry.

c   Which country has the lower opportunity cost in the production of lorries?

(Use Guided question 2 on opportunity cost to help you with this question.)

7   The ratio of dependents to the working-age population in an economy is $1:2$. This means that for every dependent in the population, there are two people of working age.

a   What proportion of the total population are dependents and what proportion are of working age?

b   If the population of the economy is 60 million, calculate the number of dependents.

2 Percentages, percentage changes and elasticity

Percentages

A percentage is a number expressed as a fraction of 100. The percentage symbol % means 'divided by 100' or 'out of 100'. So, for example, 40% can be expressed as $\frac{40}{100}$ and vice versa.

To convert a fraction to a percentage, calculate the fraction in decimal form and multiply by 100. For example, two parts out of a total of five can be converted to a percentage as follows:

$$\frac{2}{5} \times 100 = 0.4 \times 100 = 40\%$$

Just like finding a fraction or proportion of a total amount, you can also find a percentage of a quantity. For example, to find 40% of £270, first write 40% as a fraction or as a decimal and then multiply it by £270:

$$40\% = \frac{40}{100}$$

$$\frac{40}{100} \times £270 = £108$$

Percentages are used in a variety of economic situations, including:

- **taxation** — for example, VAT is currently set at 20% of the price of a good or service
- **interest rates** — these are expressed in percentage terms, such as the base rate set by the Bank of England at 0.5% in 2015
- **employment and unemployment** — the rates of employment and unemployment are usually given as percentages of the working population
- **costs, revenues and profits** — businesses often find it useful to calculate, for example, the percentage of their costs allocated to advertising or the percentage of revenue earned in different areas of a country
- **market share and concentration ratios** — market share is the percentage of total revenue a firm earns in that market. Market share figures can be used to calculate concentration ratios, which show the amount of market share controlled by the top firms in a market

A Worked examples

a **In a town there are 400 000 people in the working population but 25 000 are currently unemployed. Calculate the percentage of the working population who are unemployed.**

Divide the number of unemployed workers (25 000) by the total working population (400 000) and then multiply by 100:

$$\frac{25000}{400000} \times 100$$
$$= 0.0625 \times 100$$
$$= 6.25$$

Therefore, the percentage of the working population who are unemployed is 6.25%.

b In 2015, total takeaway sales revenue in a seaside town was £30 500. One fish and chip shop managed to gain 45% market share. Calculate the sales revenue of this fish and chip shop in 2015.

You are given the percentage of sales revenue (i.e. market share) for the fish and chip shop, and you can use this to work out its sales revenue in 2015.

Convert 45% into a fraction of 100 and then multiply it by the total takeaway sales revenue in the town (£30 500):

$$45\% = \frac{45}{100}$$

$$\frac{45}{100} \times £30\,500 = £13\,725$$

B Guided questions

Copy out the workings and complete the answers on a separate piece of paper.

1 **The top four firms in a market have market shares of 45%, 20%, 15% and 4%. The other firms have 16% of the market. Calculate the three-firm concentration ratio for this market.**

- A concentration ratio shows how much market share the top firms have in that market.
- A three-firm concentration ratio means the market share of the top three firms in the market.
- Although 'other firms' have 16% of the market, which seems higher than the 15% market share of the third firm, be careful not to include the 'other firms' figure in the three-firm concentration ratio, because it is made up of multiple firms, not a single firm.

Add the market shares of the top three firms together.

45% + _____% + _____%

= _____%

2 **Mitchell put £2500 into his savings account at the start of the year. His bank manager has offered him 2% annual interest, paid at the end of the year. Calculate Mitchell's bank balance at the end of the year.**

Step 1: convert 2% to a fraction of 100 and multiply it by Mitchell's deposit at the start of the year. This gives the amount of interest that will be paid at the end of the year.

$$\frac{2}{100} \times £2500 = £ \,\rule{2cm}{0.4pt}$$

Step 2: add the amount of interest to Mitchell's original deposit to find his final bank balance.

£ _____ + £2500 = £ _____

3 According to HM Treasury, in 2015–16 the government will spend £743 billion in total. Some key areas of spending are shown in Table 2.1.

Table 2.1 Government spending (2015–16)

Area	Spending
Education	£99 billion
Healthcare	£141 billion
Defence	£45 billion
Social protection	£232 billion

Giving your answers to three significant figures (3 s.f.), calculate the percentage of government spending devoted to:

a education

Select the 'education' figure from the table and divide it by total government spending. Then multiply by 100. Round your final answer to three significant figures.

$$\frac{£99 \text{ billion}}{£743 \text{ billion}} \times 100 = \underline{\hspace{2cm}}\%$$

b social protection

Repeat the process for social protection.

4 Catherine earned £14 650 from her job this year and is calculating how much income tax she needs to pay. She will pay no tax on the first £10 000 of her income, while the rest of her income is taxed at 20%. Calculate:

a how much income tax Catherine needs to pay to the government

Catherine will pay tax only on income above £10 000, so £4650 of her income will be taxed at 20%.

$$\text{Income tax} = \frac{}{100} \times (£14\,650 - £10\,000)$$

$$= £ \underline{\hspace{2cm}}$$

b the percentage of her income that will be paid in tax

Find the percentage of her total income that your answer to part **a** accounts for.

C Practice questions

5 A building company calculated that its total costs in 2015 were £1.2 million. Of this amount, it spent £300 000 on wages and salaries. What percentage of total costs was spent on wages and salaries?

6 A government raised £620 billion in tax revenue in 2014–15, and 30% of this came from corporation tax. Calculate the total amount of corporation tax revenue raised in 2014–15.

7 A country's GDP is £745 billion. In the current financial year it ran a budget deficit of £54 billion and its national debt is £612 billion. Calculate, as a percentage of GDP, the country's:

a budget deficit

b national debt

8 In a certain country there are 4.5 million people in the working-age population.

Table 2.2

Percentage of working-age population	
Unemployed	6
Employed	72
Inactive	22

According to the data in Table 2.2, calculate the number of working-age people in the country who are:

a employed

b unemployed

9 In a large town there are four cinemas operating. Table 2.3 shows the total revenue of each cinema in the first three months of 2015.

Table 2.3 Cinema revenue

Cinema	Total revenue
Watch	£330 500
The Box	£220 750
Le Public Cinéphile	£99 500
Central Cinema	£98 750

Calculate, to the nearest percentage, the three-firm concentration ratio for this market.

Percentage changes

A percentage change is a change in quantity **relative to the original amount**, expressed as a percentage. Percentage changes can give you an idea of how much variables have changed in comparison to their original amounts, which may not be as clear when looking at absolute figures.

To calculate a percentage change, divide the change in the variable by the original amount and multiply by 100:

$$\text{percentage change} = \frac{(\text{new value} - \text{original value})}{\text{original value}} \times 100$$

Sometimes it will be useful to calculate the original value or the new value from known percentage change figures. You may find it helpful to learn the following formulae to help you with related problems on percentage changes:

$$\text{new value} = \text{original value} \times \left(1 + \frac{\text{percentage change}}{100}\right)$$

$$\text{original value} = \text{new value} \div \left(1 + \frac{\text{percentage change}}{100}\right)$$

Percentage changes are used in many economic situations. Here are a few examples:

- **economic growth** — this is calculated as the percentage increase in GDP of an economy
- **inflation** — this refers to a general sustained increase in the price level in an economy, and is often expressed as the percentage increase in a price index, such as the consumer price index
- **price changes** — trends in share, commodity and house prices can be analysed by looking at percentage changes over time

A **Worked examples**

a   **The sterling to US dollar exchange rate has increased from £1 : $1.40 to £1 : $1.50. Calculate the percentage increase in the sterling exchange rate against the dollar.**

Use the percentage change formula given above and substitute in the relevant numbers:

$$\text{percentage change} = \frac{(\text{new value} - \text{original value})}{\text{original value}} \times 100$$

$$= \frac{(\$1.50 - \$1.40)}{\$1.40} \times 100 = \frac{\$0.10}{\$1.40} \times 100$$

$$= 7.14285\ldots$$

It is a good idea to round your answer even if you are not specifically asked to do so. Rounding to three significant figures, the percentage increase in the exchange rate is 7.14%.

b   **An investor is considering whether to buy or sell shares in BP and easyJet. Table 2.4 shows information about the share prices of the two companies at the start and end of the previous trading day, as well as the percentage changes over that day. Calculate the values of A and B in the table.**

Table 2.4

	Start of trading	End of trading	Percentage change
BP	£460.00	A	Down 5%
easyJet	B	£1 605.90	Up 1%

Step 1: the value of A is the 'new value' of BP shares at the end of the trading day, so use the formula

$$\text{new value} = \text{original value} \times \left(1 + \frac{\text{percentage change}}{100}\right)$$

Substitute the known figures for BP into the formula. Note that 'down 5%' means that the percentage change has a negative sign, so use '−5' in the formula:

$$\text{new value} = 460 \times \left(1 + \frac{-5}{100}\right)$$

$$= 460 \times (1 - 0.05) = 460 \times 0.95$$

$$= 437$$

Therefore A = £437.00.

Step 2: the value of B is the 'original value' of easyJet shares at the beginning of the trading day, so use the formula

$$\text{original value} = \text{new value} \div \left(1 + \frac{\text{percentage change}}{100}\right)$$

Substitute the known figures for easyJet into the formula:

$$\text{original value} = 1605.9 \div \left(1 + \frac{1}{100}\right)$$
$$= 1605.9 \div (1 + 0.01)$$
$$= 1605.9 \div 1.01$$
$$= 1590$$

Therefore B = £1590.00.

B Guided questions

Copy out the workings and complete the answers on a separate piece of paper.

1 **The real GDP of one European economy rose from €1300 billion in 2013 to €1352 billion in 2014. Calculate the rate of economic growth in 2014.**

The rate of economic growth is the percentage increase in real GDP over time. So use the percentage change formula with the given real GDP figures:

- new value = _____
- original value = _____

$$\text{percentage change} = \frac{(\text{new value} - \text{original value})}{\text{original value}} \times 100$$

$$= \underline{\hspace{6cm}}$$

2 **Oil prices are currently \$72 a barrel following a 40% reduction over the last 6 months. What was the oil price 6 months ago?**

The oil price 6 months ago is the 'original price' prior to the reduction. So use the formula

$$\text{original value} = \text{new value} \div \left(1 + \frac{\text{percentage change}}{100}\right)$$

with the following numbers substituted in:

- new value = _____
- percentage change = _____

Remember that 'reduction' means that the percentage change has a negative sign.

3 **Table 2.5 shows the consumer price index in February 2013 and February 2014 in the UK and Eurozone, where 2005 is the base year for both countries.**

Table 2.5

	Feb 2013	Feb 2014	Inflation rate
UK	125.2	127.4	X
Eurozone	116.1	Y	0.7%

Source: ONS and Trading Economics

a **By calculating the value of X, find the percentage point difference between the annual inflation rates of the UK and Eurozone in February 2014.**

Step 1: X is the UK inflation rate for the year up to February 2014. It can be calculated as the percentage change in the consumer price index over the year:

$$\text{percentage change} = \frac{(\text{new value} - \text{original value})}{\text{original value}} \times 100$$

Step 2: to find the percentage point difference between the annual inflation rates of the UK and Eurozone, subtract one inflation rate from the other.

X − 0.7 = _____ − 0.7 = _____

b **Calculate the value of Y, giving your answer to one decimal place.**

For the Eurozone, you know the 'original' value of the consumer price index in February 2013 and also the percentage change in the price index (i.e. the inflation rate) over the subsequent year, so you can use these values to calculate the 'new' price index in February 2014.

C Practice questions

4 A company made a £45 million profit in 2013 and announced a 7.5% increase in profit the next year. How much profit did it make in 2014?

5 The UK consumer price index was 127.5 in December 2013 and 128.2 in December 2014. What was the UK rate of inflation for the year to December 2014? Give your answer to two decimal places.

6 Average house prices are currently £193 048 in the UK, following an annual increase of 5.2%. How much was the average UK house worth a year earlier? Give your answer to the nearest pound.

7 In October 2015, the national minimum wage for adults rose by 20p an hour to £6.70 per hour. Calculate the percentage increase in the national minimum wage for adults.

8 Table 2.6 shows various economic statistics for an economy in 2013 and 2014.

Table 2.6

	2013	2014
Real GDP	€1 250 billion	€1 275 billion
Unemployment	1.8 million	1.35 million
Inflation	2.2%	1.5%
Budget deficit	€37.50 billion	€37.95 billion
Exchange rate	€1 : £1.20	€1 : £1.30

a Calculate the percentage change between 2013 and 2014 in:

  i unemployment

  ii the exchange rate

b What is the rate of economic growth in 2014?

9 A share in BT is currently worth £450. A number of economic forecasters were asked to predict the value of the share in 2 years' time, and their estimates ranged from a fall of 2% to a rise of 8%.

a What range of values did the forecasters predict for the BT share?

b If after 2 years the value of a BT share were to become £444, what would be the percentage change in the BT share value? Give your answer to three significant figures.

10 Table 2.7 shows changes in the price of gold (per ounce) over 3 months.

Table 2.7 Gold prices

May 2015	June 2015	July 2015
$1 210	$1 216	$1 223

a By calculating percentage changes, determine whether the monthly percentage change in gold prices was higher in June or July of 2015.

b What is the percentage point change in gold prices between June 2015 and July 2015?

c It is predicted that gold prices will be 12% lower at the end of the year compared to July 2015. What is the forecast value of gold at the end of 2015?

Elasticity calculations

In economics, **elasticity** measures the responsiveness of one variable in relation to a change in another variable. There are four elasticity concepts that you need to know for the AS/A-level course. These concepts, their standard abbreviations and the formulae for calculating them are summarised in Table 2.8. Notice that all of these concepts are defined in terms of **percentage changes**, often abbreviated '%Δ', where the Greek letter Δ (delta) stands for 'change in'.

Table 2.8 Formulae for calculating elasticities

	% change		% change
Price elasticity of demand (PED)	Quantity demanded	÷	Price
Income elasticity of demand (YED)	Quantity demanded	÷	(Real) Income
Cross-price elasticity of demand (XED)	Quantity demanded of good X	÷	Price of good Y
Price elasticity of supply (PES)	Quantity supplied	÷	Price

You also need to be able to interpret what different values of the four elasticity concepts mean. A summary of the key interpretations is given in the following table.

Table 2.9 Interpretation of elasticity values

	Positive	Negative	Responsive	Unresponsive
Price elasticity of demand (PED)	Never	Always	Price-elastic demand (−1 or below)	Price-inelastic demand (between 0 and −1)
Income elasticity of demand (YED)	Normal good (can be a luxury if the value is high, such as +3, and necessity is low, such as +0.5)	Inferior good	Income-elastic demand (above +1 or below −1)	Income-inelastic demand (between −1 and +1)
Cross-price elasticity of demand (XED)	Substitutes	Complements	Stronger substitutes or complements (higher values)	Weaker substitutes or complements (lower values)
Price elasticity of supply (PES)	Always	Never	Price-elastic supply (+1 or above)	Price-inelastic supply (between 0 and +1)

Ⓐ Worked examples

a   **A bakery increased the price of a loaf of bread from 50p to 75p. The quantity demanded dropped from 150 loaves to 135 loaves sold.**

i   **Calculate the PED of bread.**

Step 1: to calculate the PED, first find the two values in the formula — the percentage change in quantity demanded and the percentage change in price.

$$\%\Delta \text{ quantity demanded} = \frac{(\text{new value} - \text{original value})}{\text{original value}} \times 100\%$$

$$= \frac{(135 - 150)}{150} \times 100\%$$

$$= -10\%$$

$$\%\Delta \text{ price} = \frac{(\text{new value} - \text{original value})}{\text{original value}} \times 100\%$$

$$= \frac{(75\text{p} - 50\text{p})}{50\text{p}} \times 100\%$$

$$= 50\%$$

Step 2: substitute these values into the PED formula.

$$\text{PED} = \frac{\%\Delta \text{ quantity demanded}}{\%\Delta \text{ price}}$$

$$= \frac{-10\%}{50\%} = -0.2$$

ii Calculate the total revenue after the price increase.

The total revenue is the price per unit multiplied by the quantity sold.
Here it may be easier to work in pounds rather than pence:

$$\text{revenue} = \text{price} \times \text{quantity} = £0.75 \times 135 = £101.25$$

It is common for PED questions to involve total revenue calculations.

b Real incomes in the UK have risen by 0.5% this year. Table 2.10 shows the change in quantity demanded and YED figures for two holiday destinations of Britons.

Table 2.10

Destination	% change in quantity demanded	YED
Cornwall	0.25% decrease	A
Barcelona	B	+1.5

i Calculate the values of A and B.

■ The unknown amount A is a YED value, so use the standard YED formula. Remember that '0.25% decrease' means that the percentage change has a negative sign.

$$A = YED = \frac{\%\Delta \text{ quantity demanded}}{\%\Delta \text{ income}}$$

$$= \frac{-0.25\%}{0.5\%} = -0.5$$

■ To find B, which is a percentage change in quantity demanded, first rearrange the YED formula to

$$\%\Delta \text{ quality demanded} = YED \times \%\Delta \text{ income}$$

Then

$$B = (+1.5) \times (+0.5\%) = 0.75\% \text{ increase}$$

ii Are the two holiday destinations in Table 2.10 normal or inferior goods? Explain your answer.

Cornwall is an inferior good because it has a negative YED.
Barcelona is a normal good because it has a positive YED.

B Guided questions

Copy out the workings and complete the answers on a separate piece of paper.

1 A bar lowered the price of beer from £4.00 to £3.40 a pint. Following this, the demand for wine decreased by 2%.

a Calculate the XED of beer and wine.

Here wine is 'good X' and beer is 'good Y'.

Step 1: you are given the percentage change in demand for wine, so to use the XED formula, first calculate the percentage change in the price of beer.

$$\%\Delta \text{ price of beer} = \frac{(\text{new value} - \text{original value})}{\text{original value}} \times 100\%$$

$$= \underline{\hspace{5cm}}$$

Step 2: then use this value in the XED formula:

$$XED = \frac{\%\Delta \text{ quantity demanded (good X)}}{\%\Delta \text{ price (good Y)}}$$

b **According to your calculation, are beer and wine substitutes or complements?**

Substitutes have positive XED and complements have negative XED (see Table 2.9).

Therefore beer and wine are _____

2 **The price elasticity of supply for copper is estimated to be +0.4. Following a rise in the price per ounce of copper from \$125 to \$175, what percentage increase in quantity of copper supplied would be expected?**

The price elasticity of supply is given by the formula

$$PES = \frac{\%\Delta \text{ quantity supplied}}{\%\Delta \text{ price}}$$

Step 1: the question asks for the percentage increase in quantity supplied, so rearrange the formula to

$$\%\Delta \text{ quantity supplied} = PES \times \%\Delta \text{ price}$$

Step 2: the PES value is given, so you need to find the percentage change in the price of copper:

$$\%\Delta \text{ price of copper} = \frac{(\text{new value} - \text{original value})}{\text{original value}} \times 100\%$$

$$= \underline{\hspace{4cm}}$$

Step 3: substitute this figure into the formula:

$$\%\Delta \text{ quantity supplied} = PES \times \%\Delta \text{ price}$$

$$= (+0.4) \times \underline{\hspace{2cm}} = \underline{\hspace{2cm}}$$

3 **Table 2.11 shows information about the price elasticity of demand for three types of takeaway shop in a town.**

Table 2.11

Type of takeaway	%Δ quantity demanded	%Δ price	PED
Fish and chips		−12%	−2.0
Chinese	+25%		−2.5
Indian	−5%	+1.5%	

a **Copy and complete the table.**

For each type of takeaway, manipulate the PED formula to find the missing value.

- Fish and chips: the missing value is %Δ quantity demanded.
  Multiply both sides of the PED formula by %Δ price to get

  $$\%\Delta \text{ quantity demanded} = PED \times \%\Delta \text{ price}$$

  Then substitute in the known values.

- Chinese: the missing value is %Δ price.
  Divide both sides of the formula above by PED to get

  %Δ price = %Δ quantity demanded ÷ PED

  Then substitute in the known values.
- Indian: the missing value is the PED, so just use the original PED formula.

b Do the takeaways have price-elastic or price-inelastic demand?

Look at the size of the values you found in part **a** and state whether they indicate price-inelastic or price-elastic demand. Use Table 2.9 to help you.

4 **The price of milk sold in a newsagent rose from £1.10 to £1.32 a pint. Following this price rise, the following happened:**
- **The quantity of cereal sold fell by 3%.**
- **The total revenue from milk sales rose from £275.00 to £316.80.**

Note: in this question you have to work with more than one type of elasticity, which is common in exam questions.

a Calculate the XED of milk and cereal.

Use the formula

$$XED = \frac{\%\Delta \text{ quantity demanded (good X)}}{\%\Delta \text{ price (good Y)}}$$

with cereal as 'good X' and milk as 'good Y'.

%Δ quantity demanded of cereal = _____

%Δ price of milk = _____

Substitute these values into the formula.

b Calculate the PED of milk.

Clearly the PED formula is needed here. But before you can apply the formula, you need to find the percentage changes in price and quantity demanded of milk.

Step 1: use the total revenue figures to calculate the quantity demanded of milk before and after the price rise.

total revenue = price × quantity demanded

so

$$\text{quantity demanded} = \frac{\text{total revenue}}{\text{price}}$$

Quantity demanded before price change = _____

Quantity demanded after price change = _____

Step 2: use the results of Step 1 to find the percentage change in quantity demanded.

$$\%\Delta \text{ quantity demanded} = \frac{(\text{new value} - \text{original value})}{\text{original value}} \times 100\%$$

Step 3: calculate the percentage change in the price of milk — you have done this in part **a**.

%Δ price of milk = _____

Step 4: now substitute the values from Steps 2 and 3 into the PED formula

$$PED = \frac{\%\Delta \text{ quantity demanded}}{\%\Delta \text{ price}}$$

ⓒ Practice questions

5   Following a 10% increase in the price of a commodity, the quantity supplied rose by 4%. Calculate the price elasticity of supply for this commodity.

6   Table 2.12 shows the changes in income and quantity demanded (QD) for two types of clothing.

Table 2.12

	% change in income	% change in QD	YED
Second-hand clothing	−15%	+13%	
Designer clothing	+20%	+100%	

a   Fill in the missing figures in the table.
b   Which type of clothing is likely to:
   i   be an inferior good?
   ii be a normal good?
   iii have income-elastic demand?

7   The price of a games console was reduced from £300 to £270. The demand for games that can be played on this console rose from 25 000 to 28 000.
a   Calculate the XED for the games console and its games.
b   State whether the games console and its games are examples of substitutes or complements.

8   Table 2.13 shows information about the price elasticity of demand for three different products. Calculate the values of X, Y and Z.

Table 2.13

	%Δ price	%Δ QD	PED
Alcohol	+3	X	−0.4
Tobacco	+12	−0.48	Y
Petrol	Z	+2	−0.25

9   A car manufacturer estimated the price elasticity of demand and supply for one of its car models. The values are shown below.

Table 2.14

PED	−1.5
PES	+0.8

How much will the quantity demanded *and* quantity supplied change if the price of this model of car were to:

a rise by 15%?

b fall by 20%?

10 Table 2.15 shows data about transport in a UK city. Use the information to answer the following questions.

Table 2.15

Bus journeys	YED = +0.75
Train journeys	YED = +0.4
Bus and train journeys	XED = +0.2

a Calculate the percentage change in quantity demanded of train journeys if:

   i the price of bus journeys rose by 20%

   ii incomes fell by 4%

b The daily number of bus journeys rose from 18 500 to 19 240. What percentage change in income would have caused this?

11 Copy and complete Table 2.16 showing information on income elasticity of demand for a variety of products in a town during one week.

Table 2.16

	% change in income	Original quantity	New quantity	YED
Pasta	+3%	150 bags		+0.02
Cars	+5%		30 cars	+1.60
Blu-ray discs		20 discs	25 discs	+1.20
Value ready meals	−20%	1 500 meals	1 875 meals	

12 A music shop sold 20 pianos last year at a price of £700 each. The manager estimated that the price elasticity of demand for pianos was −3.0.

a Following a 10% drop in price, calculate the estimated:

   i new quantity demanded of pianos

   ii change in total sales revenue from pianos

b Demand for guitars in store fell from 200 to 150 after the drop in price of pianos. The guitars are sold at £120 each. Calculate:

   i the XED for guitars and pianos

   ii the change in total sales revenue from guitars

c Did the total revenue from pianos and guitars together rise or fall after the reduction in price of pianos?

3 Averages and quantiles

In studying the following economics topics it will be particularly useful to understand, calculate and interpret the mean, median and relevant quantiles:

- poverty and inequality
- living standards
- labour markets

The first two topics involve analysis of income and wealth distribution data, both between countries and over time. In the third topic, comparisons of the wages of different types of employees (for example, by age or gender), both within an industry and across different industries, are often presented in terms of averages and quantiles.

Mean and median

An average represents the 'typical' value in a set of data. The two types of average that you need to know for AS/A-level economics are:

- the mean
- the median

These are the most frequently used kinds of average. If a discussion of data just refers to the 'average' without specifying which type, then it is assumed to be the mean.

Mean

To find the mean of a set of values, add up all the values and then divide by the total number of observations:

$$\text{mean} = \frac{\text{sum of all the values}}{\text{number of observations}}$$

The mean is used in a variety of economics contexts. Examples include:

- **analyses of the labour market** — for example, comparing average wage rates between industries, genders, age groups etc.
- **measures of economic performance** — for example, average unemployment rate over a period, average inflation rate, average income within an economy etc.

Ⓐ Worked examples

a **Three families are comparing notes on how much money they spend per week on food.**

Family 1	£140
Family 2	£100
Family 3	£90

What is the mean amount of money that these families spend on food each week?

- The total number of observations is 3
- The sum of the values is 140 + 100 + 90 = 330

  Therefore:

- Mean = 330 ÷ 3 = 110

So the mean amount of money that the families spend on the weekly food shop is £110.

One way of thinking about the mean is levelling out the numbers in the data set. In this case,

$$£140 + £100 + £90 = £330 = 3 \times £110$$

i.e. sharing out the total equally gives the mean.

b **Table 3.1 shows the inflation rates for three consecutive years. Calculate the mean inflation rate over this period.**

Table 3.1

Year	2012	2013	2014
Inflation rate (%)	1	−0.5	0.7

Notice that one of the inflation rates is negative. When adding up the values, remember that adding a number with a negative sign is the same as subtracting the number.

- The number of observations is 3
- The sum of the inflation rates is

  $$1 + (-0.5) + 0.7 = 1 - 0.5 + 0.7 = 1.2$$

- Mean = 1.2 ÷ 3 = 0.4

So the mean inflation rate for the period 2012–14 is 0.4%.

If you think about levelling out the numbers in the data set:

$$1 - 0.5 + 0.7 = 1.2 = 3 \times 0.4$$

It is interesting to note that in this example the mean figure, which is a positive value, can be misleading for analysts since there was actually deflation (a negative inflation rate) in 2013. Deflation is usually rare.

Median

If all the values in a data set are listed in order from the smallest to the largest, the median is the value in the centre.

To work out the median:

- Write the data values in order of size from smallest to biggest.
- Count the number of observations.
- If there is an **odd** number of observations, the median is the middle value.
- If there is an **even** number of observations, the median is halfway between the two middle values (i.e. it is the mean of the two middle values, obtained by adding them and then dividing by 2).

Full worked solutions at www.hoddereducation.co.uk/essentialmathsanswers

A Worked examples

a The following values are ticket prices for a local theatre. Find the median theatre ticket price.

£8	£53	£36	£48	£6	£22	£19

Step 1: write the values in order of increasing size.

6	8	19	22	36	48	53

Step 2: count the number of observations.

There are 7 data values, an **odd** number.

Step 3: identify the middle value(s) and find the median.

Among 7 data values, the middle value is the 4th one (with 3 values to the left and 3 values to the right), and this is 22.

So the median ticket price is £22.

b The following data shows the quantity demanded per week for a local firm, over its first 6 weeks of business. Find the median quantity demanded per week over this time period.

5	30	29	52	50	54

Step 1: write the values in order of increasing size.

5	29	30	50	52	54

Step 2: count the number of observations.

There are 6 data values, an **even** number.

Step 3: identify the middle value(s) and find the median.

The middle values are the 3rd and 4th, which are 30 and 50.

The median is halfway between these two:

$$\text{median} = 30 + \frac{50-30}{2} = 40$$

(i.e. add half the difference between the two numbers to the smaller number), or

$$\text{median} = \frac{30+50}{2} = 40$$

(i.e. calculate the mean of the two middle numbers).

So the median quantity demanded over the 6 weeks is 40 per week.

B Guided questions

Copy out the workings and complete the answers on a separate piece of paper.

1 Ten students, who all have part-time Saturday jobs, were asked how much they earn per hour. The results are recorded in Table 3.2.

Table 3.2

Student	1	2	3	4	5	6	7	8	9	10
Wage rate (£)	4	6	4	4.50	9	15	7	5	5	5.50

a **Work out the mean wage rate for these students.**

Step 1: this set of data has _____ observations.

Step 2: add up all the values.

Sum of the data values = _____

Step 3: divide the sum of the data values by the number of observations to find the mean.

Mean = _____ ÷ _____ = _____

b **Work out the median wage rate for these students.**

Step 1: write out the wage rates from the smallest to the biggest.

4, _____, _____, _____, _____, _____, _____, _____, _____, _____

Step 2: count the number of observations.

There are _____ observations. Is this an even or odd number?

Step 3: identify the middle value(s).

What are the two wage rates in the middle?

Step 4: calculate the number halfway between the two middle values (see Worked example **b** on page 31 for two ways to do this). This number will be the median.

2 **The growth rate per annum (p.a.) for a country has been recorded over a 10-year period. Table 3.3 shows the data. Calculate the average growth rate over this period.**

Table 3.3

Year	2004	2005	2006	2007	2008	2009	2010	2011	2012	2013
Growth rate p.a. (%)	6.5	7.0	6.5	2.0	−3.5	−5.0	−2.5	1.5	4.0	4.0

If a question just asks for 'the average', without specifying which type, it is assumed to be the mean.

Step 1: this set of data has _____ observations.

Step 2: add up all the values.

There are some negative growth rates. Remember that adding a number with a negative sign is the same as subtracting the number.

Sum of the data values = _____

Step 3: divide the sum of the data values by the number of observations to find the mean.

Mean = _____ ÷ _____ = _____

Full worked solutions at **www.hoddereducation.co.uk/essentialmathsanswers**

In question 1 you will notice that the mean is considerably higher than the median. It is an example of how the mean may give a misleading picture of the 'typical' value of the data when the data set contains extreme (i.e. exceptionally high or low) values.

In this case, although 7 out of the 10 students receive a hourly wage in the £4–£6 range, the mean is above this range, at £6.50. This is because an exceptionally high value of £15 is pushing up the mean. The median, however, is not affected by extreme values on either end, since it focuses on the middle of the data set. The median here is £5.25 and provides a more representative value of the typical wage rate.

- The mean will be greater than the median if the data is spread out with more extreme high values than extreme low values.
- The mean will be less than the median if the data is spread out with more extreme low values than extreme high values.

C Practice questions

3 Five graduates share a house. The annual salaries of the group are as follows:

£40 000, £20 000, £25 000, £28 000, £30 000

a Work out the mean salary.
b Work out the median salary.

4 Table 3.4 shows data on the income before tax of two different age groups of taxpayers in the UK, for the year 2012–13.

Table 3.4

	Median income before tax	Mean income before tax
Ages 20–24	£14 500	£16 400
Ages 50–54	£24 500	£35 800

Source: Gov.uk, National Statistics, adapted from 'Distribution of median and mean income and tax by age range and gender 2012–2013'

The mean income before tax is higher than the median income before tax for both age groups. Briefly explain what would cause this difference.

5 Table 3.5 shows the current account balance as a percentage of GDP for Germany and the UK.

Table 3.5 Current account balance (% of GDP)

	2011	2012	2013
Germany	6.6	7.3	7.4
UK	−1.4	−3.6	−4.2

Source: OECD.Stat, Key Short-Term Economic Indicators: Current Account % of GDP (http://stats.oecd.org/index.aspx?queryid=21764)

a Work out the mean current account balance for Germany, as a percentage of GDP, over the period 2011–13.

b Work out the mean current account balance for the UK, as a percentage of GDP, over the period 2011–13.

c Why is one mean positive and the other negative?

6 In the UK, relative poverty is defined as the proportion of people who are living on less than 60% of the median income. The median household income, after housing costs, was £374 per week in 2012–13. The Office for National Statistics (ONS) estimated that in 2012–13, the proportion of individuals living in relative poverty (after housing costs) was 21%. This represented 13.2 million people out of a total population of approximately 63 million.

Explain clearly what the above statement means.

(This involves explaining what the median measures in this context. Also try to work out what 60% of the median income per week would be, which will help to explain the above statement.)

Quantiles

The topic 'Quantiles' is assessed at A-level only by AQA, Edexcel and OCR and at both AS and A-level by WJEC/Eduqas and CCEA.

When data is listed in order of size and divided into equal subgroups, the numbers that separate the different groups, and sometimes the groups themselves, are called quantiles.

The **median** is an example of a quantile. It is the middle number of a set of data values ordered by size, so it divides the data into two halves.

If the data values are ordered from smallest to largest and split into four equal groups, the three numbers separating the groups are called **quartiles**. In this case, the median separates the second and third groups and can be referred to as the second quartile. The lower (or first) quartile separates the first and second groups and divides the bottom half of the data into two halves. The upper (or third) quartile separates the third and fourth groups and divides the upper half of the data into two halves.

Data in economics is often organised by **quintiles**. This means that the data values are ordered from smallest to largest and split into five equal groups.

In economics, **deciles** are also commonly used, where the data has been ordered by size and split into 10 equal groups.

Percentiles are obtained when the data is divided into 100 equal groups. The lower quartile is equivalent to the 25th percentile, while the upper quartile is the same as the 75th percentile.

It is unlikely that exam questions will require students to calculate quantiles, but an overview of how this is done (shown in Worked example **a** and Guided question **1**) will make data presented in this way easier to understand and analyse.

Worked examples

a **Over seven working days, Susan records how late her bus is on each day. The following are the numbers of minutes her bus is late by:**

8, 8, 16, 5, 17, 10, 12

Find the median and upper and lower quartiles of this data.

'Quartiles' means that the data should be divided into four equal groups.

Step 1: sort the data values in order of size, from smallest to largest.

| 5 | 8 | 8 | 10 | 12 | 16 | 17 |

Step 2: find the median, which divides the data into two halves.

The number of observations is 7, an odd number, so the median is the middle value, which is 10.

Step 3: look at each half of the data (to the left and to the right of the median), and find the median of that half. This will give the lower and upper quartiles.

To the left of the median you have

| 5 | 8 | 8 |

The median of these three values is 8, so the lower quartile is 8.

To the right of the median you have

| 12 | 16 | 17 |

The median of these three values is 16, so the upper quartile is 16.

To summarise, you have

Lower quartile	Median	Upper quartile
$\frac{1}{4}$	$\frac{1}{2}$	$\frac{3}{4}$
[5 8 8]	10	[12 16 17]

b **The following is data on total household income for Great Britain in 2010–12.**

	Lower quartile (25th percentile)	Median (50th percentile)	Upper quartile (75th percentile)
Total household income	£18 000	£32 100	£53 500

Source: ONS Statistical Bulletin, 'Wealth and Income, 2010–12' (published 3 July 2014). Data extracted from Table 1: Total household wealth and income, summary statistics: Great Britain, 2010–12.

i **Explain what is meant by 'lower quartile'.**

Households in Great Britain were ranked in order from the poorest to the richest (in terms of household income). The lower quartile is the household income a quarter of the way through the list. This is £18 000, which means that in 2010–12, a quarter (25%) of households in Great Britain had an income of £18 000 or less.

ii **What fraction of households have an income between £32 100 and £53 500?**

From the data, £32 100 is the median household income and £53 500 is the upper quartile. The median household income is the middle value when all household incomes have been listed in order from lowest to highest. It is halfway, or two-quarters of the way, through the distribution. The upper quartile is three-quarters of the way through the distribution. This means that one-quarter (25%) of households in Great Britain had an income between £32 100 and £53 500.

iii **Using the given data, comment on the distribution of household income.**

The lower quartile is £18 000, which means that a quarter (25%) of households had an income equal to or less than £18 000. The median household income is £32 100, which means that the 'middle' household income is £32 100, so a half of all households are poorer and a half are richer than this. The upper quartile is £53 500, which means that the richest quarter (25%) of households had income the same as or in excess of £53 500.

B Guided questions

Copy out the workings and complete the answers on a separate piece of paper.

1 **The population of a fictional country is 10 people. Their weekly earnings, measured in pounds, are shown below:**

400, 350, 50, 150, 250, 400, 300, 300, 200, 150

a **What is the median?**

Step 1: sort the data values in order of size, from smallest to largest.

50, _____, _____, _____, _____, _____, _____, _____, _____, _____

Step 2: find the median, which divides the data into two halves.

The number of observations is 10, an even number, so the median is halfway between the two middle values.

The two middle numbers are _____ and _____

Therefore the median is _____

b **What is the lower quartile?**

Look at the data to the left of the median, and find the median of this group of values.

c **What is the upper quartile?**

Look at the data to the right of the median, and find the median of this group of values.

2 Table 3.6 shows the distribution of disposable household income (as percentages of total household disposable income) for the UK in 1979. What share of total disposable income did the richest 40% of UK households have in 1979?

Table 3.6

	Quintile groups of all households				
	Bottom	2nd	3rd	4th	Top
% of total household disposable income	10	14	18	23	35

Source: ONS Economic and Labour Market Review (16 December 2008), 'The distribution of household income 1977 to 2006/07'. Data adapted from Figure 3: Shares of total equivalised household income by quintile group.

Step 1: identify which quintiles represent the richest 40% of households.

Step 2: write down the percentages of total household disposable income for these quintiles.

Step 3: add up the percentages in Step 2 to obtain the share of total disposable income accounted for by the richest 40% of UK households.

Note: it is interesting to compare this answer with the share of total disposable income that the poorest 40% of households have. The poorest 40% of households account for 10% + 14% = 24% of the total disposable income.

ⓒ Practice questions

3 Seven students compare the wage rates they receive from their Saturday jobs:

£4   £6   £5   £8   £10   £4.50   £5

  a   What is the median wage rate?
  b   What is the lower quartile of the wage rates?
  c   What is the upper quartile of the wage rates?

4 The following is data on total household wealth for Great Britain in 2010–12.

	Lower quartile	Median	Upper quartile
Total household wealth	£57 000	£218 400	£490 900

Source: ONS Statistical Bulletin, 'Wealth and Income, 2010–12' (published 3 July 2014). Data extracted from Table 1: Total household wealth and income, summary statistics: Great Britain, 2010–12.

  a   If households were lined up in order of increasing wealth, how much wealth would the middle household have?
  b   What percentage of households have wealth of £57 000 or less?
  c   What proportion of households have wealth between £57 000 and £490 900?

5   Table 3.7 shows the annual income before taxes paid or benefits received of quintile groups of households for a fictional country, measured in pounds.

Table 3.7

	Quintile groups of all households				
	Bottom	2nd	3rd	4th	Top
Average income per household (£)	5 000	9 000	20 000	40 000	80 000

a   The data shows **average** income per household in each quintile group. Explain what this means.

b   What is the average income of households who fall into the top 20% of household incomes in the country?

c   What is the ratio of the average income per household of the richest 20% to the average income per household of the poorest 20%?

4 Money and index numbers

Index numbers

For all exam boards, the skill of interpreting index numbers is required at both AS and A-level. The skill of calculating with index numbers is assessed at A-level only by OCR and at both AS and A-level by AQA, Edexcel, WJEC/Eduqas and CCEA.

Index numbers are used to track economic variables and make comparisons over time. They are useful in a wide range of economic situations. Here are a couple of examples:

- **GDP** — index numbers provide a quick way of showing how GDP has been changing relative to the GDP of a particular time period, such as the financial crisis.
- **Inflation** — inflation rates are calculated using index numbers such as the consumer price index.

To define index numbers for an economic variable, choose a base time period (usually a base year) and set the index for this period to 100. Then compare the values of the economic variable in other time periods to the value in the base period, using the formula

$$\text{index number} = \frac{\text{current figure}}{\text{figure in base time period}} \times 100$$

An index number allows you to quickly see the percentage change in a variable relative to its value in the base period. For example, if 2008 is the base year and in 2009 the index was 105 (which is 5 above 100), this tells you immediately that from 2008 to 2009 there had been a 5% increase in the value of the variable.

Besides using the formula above to calculate index numbers, you can also work out current and base year figures, given appropriate information, by rearranging the above formula to the following versions:

$$\text{current figure} = \text{figure in base time period} \times \frac{\text{index number}}{100}$$

$$\text{figure in base time period} = \text{current figure} \div \frac{\text{index number}}{100}$$

A Worked examples

a At a certain cinema, the price of a ticket rose from £5.00 in 2014 to £6.15 the next year.

   i Calculate an index number for the cinema ticket price in 2015, assuming that 2014 is the base year and therefore has an index equal to 100.

     The base year is 2014 and its index number is set to 100. To calculate the index number for 2015, apply the formula:

$$\text{index number} = \frac{\text{current figure}}{\text{figure in base time period}} \times 100$$

$$= \frac{£6.15}{£5.00} \times 100$$

$$= 123$$

ii **Using your answer to i, state the percentage increase in the cinema ticket price between 2014 and 2015.**

The index number for 2015, 123, is 23 above 100. This means that cinema tickets were 23% more expensive in 2015 than in 2014, i.e. the percentage increase in the cinema ticket price was 23%.

b **Table 4.1 shows data on average house prices in two regions of the UK. Using June 2013 as the base month, calculate the values of X and Y.**

Table 4.1

	Average house price June 2013	Average house price January 2015	Index number January 2015
South East	£281 000	X	120
North West	Y	£173 600	112

Source: BBC News

- The value X is an average house price in January 2015 (not the base period), so calculate it using the formula

$$\text{current figure} = \text{figure in base time period} \times \frac{\text{index number}}{100}$$

$$X = £281\,000 \times \frac{120}{100}$$

$$= £337\,200$$

- The value Y is an average house price in June 2013 (the base period), so calculate it using the formula

$$\text{figure in base time period} = \text{current figure} \div \frac{\text{index number}}{100}$$

$$Y = £173\,600 \div \frac{112}{100}$$

$$= £155\,000$$

B Guided questions

Copy out the workings and complete the answers on a separate piece of paper.

1 **Table 4.2 shows an index for GDP in a Eurozone economy after the financial crisis.**

Table 4.2

	GDP (2008 = 100)
2009	97.5
2010	95.0
2011	98.2

a **What was the country's rate of economic growth in 2009?**

- In the table, '(2008 = 100)' means that 2008 is the base time period, with the index set to 100.

- The rate of economic growth is the percentage change in GDP.
  We can work this out by comparing the index number for 2009, 97.5, with 100.
- Because 97.5 is below 100, the growth rate will be negative.

b How far below the 2008 level was the country's GDP in 2011?

Again, look at the difference between the index for 2011 and 100 (the index for the base year 2008).

2 The profits of a company were £260 million in 2012. Following this, profits were £286 million in 2013, before falling to £104 million in 2014.

Using 2012 as the base year, calculate the index numbers for the company's profits in 2012, 2013 and 2014.

- Because 2012 is the base year, the index number for 2012 is simply 100.
- For 2013 and 2014, the index numbers can be calculated using the formula

$$\text{index number} = \frac{\text{current figure}}{\text{figure in base time period}} \times 100$$

3 A statistical analyst has constructed an index for the oil price in the world economy compared to 2010, when oil prices were $70 per barrel.

Table 4.3 Index of average oil prices in comparison to 2010

Year	2011	2012	2013	2014
Index	125	123	131	122

a Calculate the highest value of oil prices between 2010 and 2014.

- 2010 is the base year, with index number 100.
- The highest oil prices were in 2013, as this year had the highest index number.
- To calculate the oil price in 2013, use the formula

$$\text{current figure} = \text{figure in base time period} \times \frac{\text{index number}}{100}$$

b The analyst forecast that oil prices would average $52.50 in 2015. Calculate the index number for 2015, assuming the forecast is correct.

To calculate the index for 2015, use the standard index number formula comparing the figure for 2015 to that for 2010.

C Practice questions

4 In 2014 the volume of exports from the UK was 12% above the 2008 level. Taking 2008 to be the base year, what is the index number for 2014?

5 Using January 2013 as the base time period, the price index for the UK economy was 102.3 in January 2014 and 96 in January 2015.
a State the rate of inflation for the UK from January 2013 to January 2014.
b By what percentage did price levels fall in the UK between January 2013 and January 2015?

6 The GDP of an economy was $435 billion in 2010 and is $470 billion this year. Using 2010 as the base year, calculate an index number for the GDP this year.

7 The current pound sterling to US dollar exchange rate is £1 : $1.45, which is down by 8% from a year earlier.

   a Taking a year ago as the base period, state the index number for the current exchange rate.

   b Using the index number from part **a** or otherwise, calculate the pound's exchange rate against the dollar a year ago.

8 Table 4.4 shows tax revenue receipts for an economy over four financial years.

Table 4.4

	2010–11	2011–12	2012–13	2013–14
Tax revenue receipts	£828 billion	£720 billion	£648 billion	£792 billion
Index number		100		

   a Using 2011–12 as the base year, calculate the index numbers for the other three financial years in the table.

   b The index number for 2014–15 was forecast to be between 102 and 104. What was the range of values forecast for tax revenue receipts in 2014–15?

9 An investor has calculated an index for the shares in her portfolio using 21 March 2014, the date on which the portfolio was created, as the base period. The portfolio is currently worth £4375 and the index currently stands at 125.

   How much value has the share portfolio gained since its creation:

   a in percentage terms?

   b in pounds?

10 Table 4.5 shows the value of an index for the price of silver at the end of the first 10 days of April compared to the start of the year (1 January = 100).

Table 4.5 Index for the price of silver in April

Day 1	Day 2	Day 3	Day 4	Day 5	Day 6	Day 7	Day 8	Day 9	Day 10
101.5	102.2	98.1	99.0	99.0	87.2	87.5	90.2	90.1	89.1

The price of silver at the start of the year was $14.50 per ounce. Calculate the price of silver at the end of:

   a 1 April

   b 10 April

Converting money to real terms

The topic 'Converting money to real terms' is assessed at A-level only by Edexcel, OCR and WJEC/Eduqas and at both AS and A-level by AQA and CCEA.

When interpreting or analysing economic statistics, it is important to note whether they are presented in nominal or real terms.

■ **Nominal** (or money) values have not been adjusted for inflation.
■ **Real** values have been adjusted for inflation.

As real values take inflation into account, they reflect the purchasing power of the currency and can give a more accurate picture of the economic situation.

For example, workers may receive an annual increase in their nominal wages. However, if there has been significant inflation, in real terms the workers may be no better off as the purchasing power of their money may have fallen.

In a number of economic contexts you may need to convert from nominal to real terms. Here are a few examples:

- **national income statistics** — for example, nominal GDP converted to real GDP
- **wages and salaries** — for example, nominal wages converted to real wages
- **savings and debts** — for example, nominal debt converted to real debt
- **commodities** — for example, nominal oil prices converted to real oil prices

Converting from nominal to real terms relies on an appropriate **price index**. The formula is

$$\text{real value} = \frac{\text{index of comparison period}}{\text{index of current period}} \times \text{nominal value}$$

The 'comparison period' could be the base period of the index, so that its index number is 100. The 'current period' is the time period for which you want to calculate the real value.

When converting from nominal to real terms, it is important to note that the final answer will depend on the time period you are comparing to. To make this clear, the comparison period is usually given when you are presented with economic data.

For example, a chart might be labelled 'GDP at constant 2005 prices'. This tells you that real GDP is being shown and that the comparison time period is 2005. Therefore the numerator (top of the fraction) in the formula will be the price index for 2005.

 Worked examples

a **Table 4.6 shows seasonally adjusted GDP and the consumer price index in the UK between 2012 and 2014. (If you want to know what 'seasonally adjusted' means, see 'Seasonally adjusted figures' on page 86 in Unit 8.)**

Table 4.6

	GDP (£ billion)	Consumer price index (2005 = 100)
2012	1 628	123
2013	1 655	126.1
2014	1 702	128

Source: ONS

i **Calculate real GDP for 2012 and 2014, using 2005 as the comparison year.**

Use the formula

$$\text{real value} = \frac{\text{index of comparison year}}{\text{index of current year}} \times \text{nominal value}$$

where 'index of comparison year' is 100, because the comparison year is 2005, the base year of the price index.

For 2012:

$$\text{real GDP} = \frac{100}{\text{index for 2012}} \times \text{nominal value for 2012}$$

$$= \frac{100}{123} \times \text{£1628 billion}$$

$$= \text{£1324 billion (to the nearest billion)}$$

For 2014:

$$\text{real GDP} = \frac{100}{\text{index for 2014}} \times \text{nominal value for 2014}$$

$$= \frac{100}{128} \times \text{£1702 billion}$$

$$= \text{£1330 billion (to the nearest billion)}$$

ii **Calculate the percentage change in real GDP from 2012 to 2014.**

Use the percentage change formula (see page 18):

$$\text{percentage change} = \frac{\left(\text{new value} - \text{original value}\right)}{\text{original value}} \times 100$$

$$= \frac{(\text{£1330 billion} - \text{£1324 billion})}{\text{£1324 billion}} \times 100$$

$$= 0.45317 \ldots$$

So from 2012 to 2014 the percentage change in real GDP was 0.45% (2 d.p.).

b **According to the ONS, in 2013 the average real income of a worker in the UK was £21 015, using 2005 as the base year with a consumer price index of 100.**

Given that the consumer price index was 126 in 2013, calculate the average income of a UK worker in nominal terms. Give your answer to the nearest pound.

In this question you need to calculate a nominal value, given the real value and the price index.

Step 1: rearrange the formula

$$\text{real value} = \frac{\text{index of comparison year}}{\text{index of current year}} \times \text{nominal value}$$

to make 'nominal value' the subject of the formula. This gives

$$\text{nominal value} = \text{real value} \times \frac{\text{index of current year}}{\text{index of comparison year}}$$

Step 2: substitute the given figures into the rearranged formula.
Therefore:

$$\text{nominal income} = \text{real income} \times \frac{\text{index of 2013}}{\text{index of 2005}}$$

$$= \text{£21 015} \times \frac{126}{100}$$

$$= \text{£26 478.90}$$

$$= \text{£26 479 (to the nearest pound)}$$

B Guided questions

Copy out the workings and complete the answers on a separate piece of paper.

1   **Following the financial crisis in 2008, an American decided to start saving and by 2014 had accumulated $25 000. The consumer price index in the USA was 237 in 2014 and 215 in 2008. Calculate the real value of the savings in 2008 dollars.**

   Step 1: decide what the 'comparison period' and the 'current period' are.

   The phrase 'real value … in 2008 dollars' means that 2008 is the comparison period. In this case, 2014 is the current period.

   Step 2: write down the formula to use.

   Use the standard conversion formula

   $$\text{real value} = \frac{\text{index of comparison period}}{\text{index of current period}} \times \text{nominal value}$$

   Step 3: substitute the appropriate figures into the formula.

   - index of comparison period = _____
   - index of current period = _____
   - nominal value = _____

2   **In March 1974, the world oil price almost quadrupled in nominal terms to $12 per barrel. An oil price of $12 would be the equivalent of $59 in the US economy in March 2014, in real terms.**

   a   **Using March 1974 as the base period, with an index of 100, calculate the price index for the US economy in March 2014.**

   Step 1: decide what the 'comparison period' and the 'current period' are.

   The phrase 'the equivalent of … in March 2014' tells us that March 2014 is the comparison period. (Note that this is not the same as the base period of the index, which is given as March 1974.)

   In this case, the 'current period', the time period for which you have the nominal and real values, is March 1974.

   Step 2: write down the formula to use.

   In this question you need to find the price index for the comparison period (March 2014), so first rearrange the standard conversion formula

   $$\text{real value} = \frac{\text{index of comparison period}}{\text{index of current period}} \times \text{nominal value}$$

   to make 'index of comparison period' the subject of the formula. This gives

   $$\text{index of comparison period} = \frac{\text{real value}}{\text{nominal value}} \times \text{index of current period}$$

Step 3: substitute the appropriate figures into the formula.

- index of current period = _____
- real value = _____
- nominal value = _____

b Calculate the rate of inflation in the US economy over the 40 years between March 1974 and March 2014.

The rate of inflation is the percentage change in the price index.
As March 1974 is the base period for the price index, you can calculate the inflation rate by simply subtracting 100 from the price index in March 2014.

3 **Table 4.7 shows nominal and real seasonally adjusted consumption figures, as well as the consumer price index, in the UK for a few recent years. Calculate the values of A and B.**

Table 4.7

	Consumption (£ billion)	Consumption, 2010 prices (£ billion)	Consumer price index (2010 = 100)
2012	1022	951.6	107.4
2013	A	961.9	110.1
2014	1103	B	111.8

Source: ONS

- The heading 'Consumption, 2010 prices' tells you that this column shows real figures and that the comparison period is 2010. The column headed simply 'Consumption' shows nominal figures.
- A is a nominal value, so to calculate it, first rearrange the standard conversion formula to make 'nominal value' the subject of the formula. Then substitute appropriate figures into the rearranged formula.
- B is a real value, so to calculate it, just use the standard conversion formula.

Ⓒ Practice questions

4 A UK business has accumulated debt in nominal terms of £8500 in 2015. Given that the UK consumer price index was 100 in 2005 and 129 in 2015, calculate the real debt of this business, assuming constant 2005 prices.

5 Read the following economic data:
- In September 2010 the gold price hit a record high of $1282.
- The US consumer price index in September 2014 was 109.0 (September 2010 = 100)

What was the real gold price in September 2010 (assuming constant September 2014 prices)?

6   Table 4.8 shows the Chinese consumer price index in three recent years.

Table 4.8

	Chinese CPI (2012 = 100)
2012	100
2013	102.6
2014	104.7

Source: World Bank

In 2014 the Chinese government planned to spend 14.9 trillion yuan in real terms, assuming constant 2013 prices. Calculate nominal Chinese government spending in 2014.

7   Using the data in Table 4.9, calculate real GDP per capita for the US economy in 2014, assuming constant 2012 prices.

Table 4.9 Consumer price index and GDP per capita in the USA

Year	Consumer price index	GDP per capita ($)
2012	230	51 457
2013	233	52 980
2014	237	54 630

Source: US Bureau of Labour Statistics and the World Bank

8   Table 4.10 shows average UK house prices between 2012 and 2014. Copy the table and complete the final column.

Table 4.10

Year	Nominal house price (£)	Real house price (£)	Consumer price index (2012 = 100)
2012	162 900	162 900	100
2013	174 000	169 800	
2014	189 000	181 600	

5 Standard graphical forms

Some of the graphs or diagrams you will encounter in your AS/A-level economics course relate to specific topics. These include PPF diagrams, supply and demand diagrams and theory of the firm diagrams. A lot of exam marks will be allocated to interpreting such diagrams, through which you will demonstrate your quantitative and analytical skills. This unit covers several standard, general graphical forms which are used across the whole course.

Graphs provide a pictorial representation of data involving two (or more) variables. A variable is a quantity whose value can change — for example, the price of a good. If data has been collected on more than one variable, a graph of the data can be drawn to show the relationship between the variables.

Representing data in graphical form can help economists identify patterns and then formulate concepts, theories and models to explain these patterns. (A model is a simplified view of the real world.) By using graphs and diagrams, economists can also analyse and test whether existing economic theories or models make sense of what is really happening in the economy.

Some advantages of presenting data in graphical form:
- The information is easier to take in at a glance, compared with data presented in a table.
- Graphs help us to identify any relationships and patterns which may exist in the data.

Some tips on interpreting graphs:
- Look at the title to see what information is being presented.
- Look at the labels of the axes to see what each axis is measuring.
- Look at the units of the data, e.g. pence, pounds, tonnes.
- Look at the scale on each axis — how big are the numbers? Some might seem small, but be aware that they could be given in thousands or millions (again, check the units).
- Take note of any source given for the data.

Main types of graphs used in economics
Scatter diagrams

A scatter diagram shows how two variables are related. The value of one variable is plotted against the corresponding value of another variable — for example, price data may be plotted against quantity demanded. This gives the usual demand curve, which is one of the first concepts you will come across in your study of economics.

Scatter diagrams are useful for identifying relationships that might exist between variables:

■ Two variables have a **positive** relationship or **direct** relationship if they move in the same direction, i.e. as one variable increases, so does the other. For example, if the price of a good rises within a certain range, the total revenue of the firm also rises.

■ Two variables have a **negative** relationship or **inverse** relationship if they move in opposite directions, i.e. as one variable increases, the other decreases. See the worked example below.

■ There is a U-shaped relationship between two variables if one of them reaches a minimum at an intermediate value of the other. For example, when average cost is plotted against output, the resulting average cost curve will be U-shaped.

■ There is an upside-down U-shaped relationship between two variables if one of them reaches a maximum at an intermediate value of the other.

■ If the points on a scatter diagram lie all over the place with no obvious pattern, there is no clear relationship between the variables.

 Worked example

Table 5.1 shows the demand for a cereal bar at different prices.

Table 5.1

Price of cereal bar (pence)	Quantity demanded per annum (000s)
45	18
48	16
50	15
60	14
65	14
72	13
75	13
80	12
85	11
90	10

i **Plot the data on a scatter diagram to show the price of cereal bars against the quantity demanded.**

Step 1: decide what information to show along each axis of the graph. In this case, the price of cereal bars should be shown on the vertical axis and the quantity demanded on the horizontal axis. Don't forget to give units with the axis labels.

Step 2: decide what scale to use for each axis. The range of values along each axis should at least cover all the data values for that variable. Look at the largest data value of each variable to determine how far the scale should go.

Step 3: plot the data on the graph:

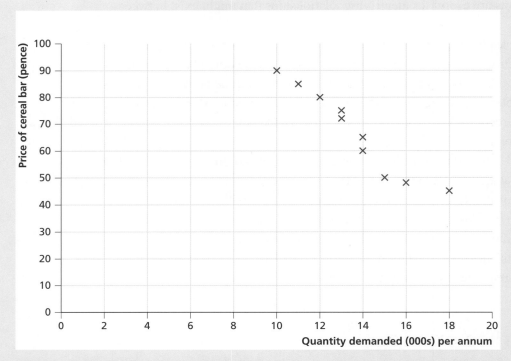

Figure 5.1 Relationship between price of cereal bar and quantity demanded per annum

ii **From the scatter diagram, what relationship between price and quantity demanded seems to exist?**

- Look at the points on the scatter diagram — could a line of best fit be drawn so that all the points lie close to the line? If so, there is a clear relationship between price and quantity demanded. Remember that a line of best fit is a line that roughly goes through the middle of the distribution of points so that there is a balance between the number of points above the line and the number of points below the line.

- Look for a general trend in the distribution of points on the scatter diagram — as you move from left to right in the diagram, do the points seem to be getting higher or lower?

- In Figure 5.1, as the price of cereal bars rises, the quantity demanded tends to fall, and vice versa. This suggests that there is an inverse relationship between the price of cereal bars and the quantity demanded (an inverse relationship exists if the two variables move in opposite directions).

- Note that the relationship is not a perfect one, where all the points are aligned on a straight line or smooth curve, which is what you often find in textbooks. In practice, other factors affecting demand may be changing at the same time, so that some data points will not fit the pattern exactly.

Time series graphs

A time series graph shows how the value of a variable is changing over time. The horizontal axis represents time and the vertical axis represents the value of the economic variable.

Time series graphs are usually line graphs, i.e. scatter diagrams where the data points have been joined up with straight lines.

Table 5.2 shows the price of a commodity over 6 months.

Table 5.2 Price per ounce of commodity X, monthly average in 2015

Month	Price of commodity (£ per ounce)
Jan	30
Feb	40
Mar	35
Apr	50
May	55
Jun	60

The following graph is constructed by plotting the prices in the middle of each month marked along the horizontal axis. Successive points are then joined up with straight lines to give a clearer picture of how prices are changing over time.

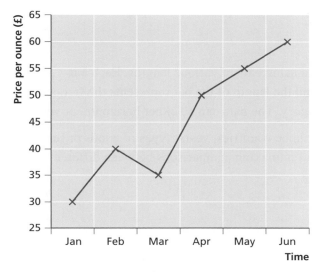

Figure 5.2 Price of commodity X

The main advantage of a time series graph is that it helps you to spot trends in the data over time. In this example, there appears to be an upward trend in the price of the commodity over time. See Unit 7 for more on trends, and Unit 8 for further analysis of time series data.

When interpreting a time series graph and using it to identify trends, be particularly careful to note the scales on the axes. The same data from Table 5.2 has been plotted as a line graph in Figure 5.3. The monthly price now seems to show much less movement, but this is only because the scale on the vertical axis is different. So make sure to look at the scale and units carefully and think about the relative size of the movements. In this case, the price per ounce has doubled by the end of the 6 months, which is a fairly significant change.

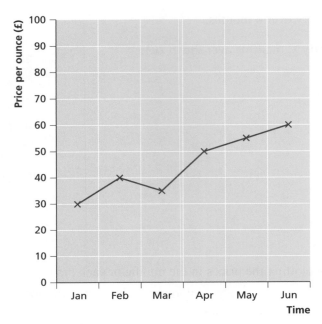

Figure 5.3 Price of commodity X

Bar charts

A bar chart is a graphical display of data using bars of various lengths. Each bar represents a distinct category of data, and the length of the bar represents the frequency of (or number of data items in) that category. The bars can be either horizontal or vertical.

In economics, the categories of data could be countries, years, types of goods or services and so on. Bar charts can be very helpful in making comparisons between different categories.

Ⓐ Worked example

Figure 5.4 shows the average annual growth of GDP per capita between 1990 and 2012 for 16 countries.

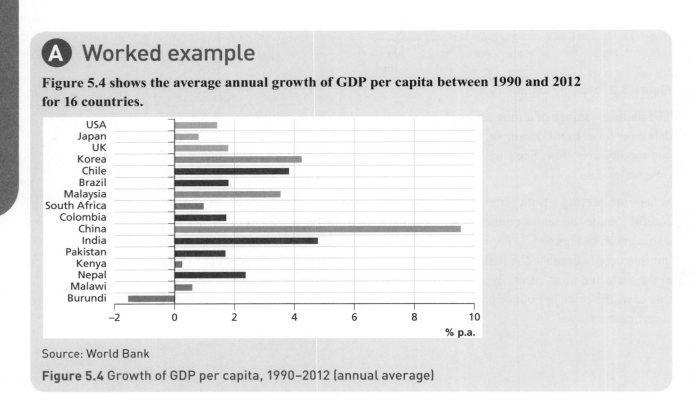

Source: World Bank

Figure 5.4 Growth of GDP per capita, 1990–2012 (annual average)

i **What was the average annual growth of GDP per capita over the period 1990–2012 for the USA?**

Place a ruler against the right end of the bar for the USA (the top bar), and look at where it hits the scale on the horizontal axis.

This shows that the average annual growth of GDP per capita for the USA was approximately 1.5%.

ii **Which country experienced the highest average annual growth in this period?**

Look at which country the longest bar corresponds to.

China had the highest average annual growth in the period 1990–2012: its average annual growth rate was approximately 9.5%.

iii **What is significant about Burundi's average annual growth rate?**

Burundi (the bottom bar) is the only country that had a negative average annual growth rate over the period 1990–2012.

Pie charts

A pie chart is so called because of its circular shape. Each 'slice' of the pie represents a category of data, and the size of the slice shows the relative frequency of that category.

The size of each slice is determined by its angle at the centre of the circle. To calculate this angle, work out the fraction (or proportion) of the total that the category represents and then multiply by 360 (as there are 360 degrees in a full circle). Guided question 2 on page 55 takes you step by step through such a calculation.

The pie chart in Figure 5.5 breaks down the different types of costs that make up the total cost of an easyJet flight. The biggest categories of cost are 'Administration' and 'Airport', which have the biggest slices of the pie (12% each).

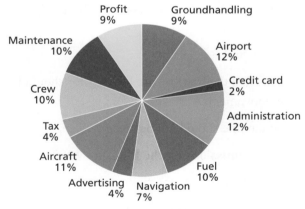

Source: Guardian, 20 August 2003

Figure 5.5 The cost of an easyJet flight

Make sure that you are familiar with and confident in interpreting all of these standard graphical forms, as they will be used to present data throughout the AS/A-level economics course. For example:

- Time series graphs are used to track macro-economic performance indicators over time.
- Pie charts are used to show the proportions accounted for by various components of government spending and tax revenue in different fiscal years.
- Bar charts are used to compare GDP per head of different countries.
- Scatter diagrams are often used in micro-economics, for instance, to illustrate the relationship between quantity demanded/quantity supplied and prices.

B Guided questions

Copy out the workings and complete the answers on a separate piece of paper.

1 Figure 5.6 shows the carbon dioxide (CO_2) emissions of four economies over a 50-year period.

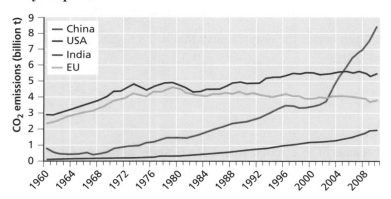

Source: World Bank

Figure 5.6 Carbon dioxide emissions

a **Which country has experienced the most rapid increase in CO_2 emissions each year since 2002?**

Look at the slope of the lines to the right of 2002. Which one is rising most steeply?

b **Since 1976, which country has experienced a slow but consistently steady rise in the level of CO_2 emissions?**

Step 1: look at the lines to the right of 1976.

Step 2: 'consistently steady rise' means that the line you are looking for should be rising without any significant erratic movements.

Step 3: 'slow' means that the slope of the line should not be steep.

c **What has happened to the level of CO_2 emissions by the EU since 1980?**

Step 1: look at the general direction of the EU line to the right of 1980 — this is the 'trend'.

Step 2: identify any periods in which movements seem more erratic (there may be none).

Step 3: use your observations from Steps 1 and 2 to write a sentence to answer the question.

2 **A firm makes four products: A, B, C and D. The sales revenue generated by each of the products in 2014 was**

A: £120 000   **B: £180 000**   **C: £60 000**   **D: £140 000**

Construct a pie chart to show the relative sales revenue contribution of each of the products towards total sales revenue for the firm.

Step 1: calculate the total sales revenue by adding up the sales revenue generated by each of the four products.

£120 000 + £ _____ + £ _____ + £ _____ = £ _____

Step 2: work out the fraction of total sales revenue that each product contributes.

Product A: $\dfrac{\text{£120 000}}{\text{total sales revenue}} = \dfrac{6}{25}$

Product B: $\dfrac{\text{£180 000}}{\text{total sales revenue}} =$ _____

Product C: $\dfrac{\text{£60 000}}{\text{total sales revenue}} =$ _____

Product D: $\dfrac{\text{£140 000}}{\text{total sales revenue}} =$ _____

Step 3: multiply each fraction by 360° to calculate the angle of each slice of the pie.

Angle of pie slice for A $= \dfrac{6}{25} \times 360° = 0.24 \times 360° = 86.4°$

Angle of pie slice for B = _____

Angle of pie slice for C = _____

Angle of pie slice for D = _____

Check that the four angles add up to 360° (but be aware that in practice the sum may differ slightly from 360° because of rounding issues).

Step 4: use the angles calculated in Step 3 to construct the pie chart.
Be sure to label each slice with the category it represents.

C Practice questions

3 Figure 5.7 shows the percentage change in real GDP (per annum) for the period 1984–93. The horizontal line represents the average growth rate during this period.

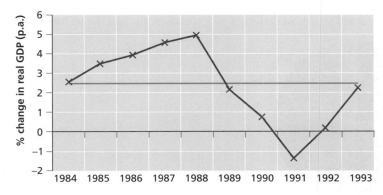

Source: ONS

Figure 5.7 A classic business cycle

a What was the percentage change in real GDP in 1988?

b What was the percentage change in real GDP in 1991?

c The average growth rate for this period was approximately 2.5%. How would this have been calculated?

4 Figure 5.8 shows the unemployment rate in January–March 2015 and the percentage change in unemployment rate on the previous quarter for different regions of the UK.

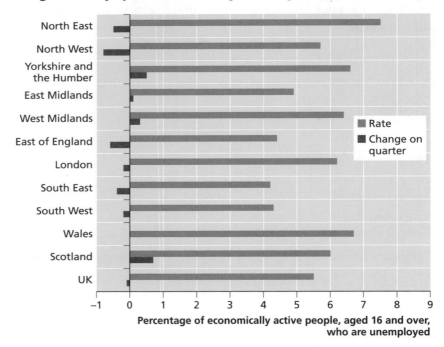

Source: ONS Statistical Bulletin: Regional Labour Market, May 2015

Figure 5.8 Unemployment rates by region, January–March 2015, seasonally adjusted

a What was the unemployment rate across the UK as a whole, for January–March 2015?

b Which region had the highest unemployment rate in this period?

c What was the unemployment rate in the South East over the period shown?

d How does the answer to part **c** compare with other regional unemployment rates during this period?

e What has happened to Scotland's unemployment rate compared with the previous quarter?

f What has happened to the unemployment rate in the majority of UK regions in the period January–March 2015 compared with the previous quarter?

5 Figure 5.9 shows the market shares of different supermarkets in the UK, in January 2003.

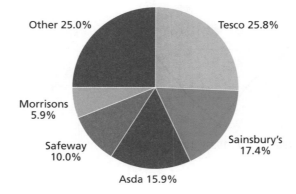

Figure 5.9 Market shares of British supermarkets, January 2003

a Which supermarket had the biggest market share in January 2003?

b The market share for Tesco in the 12 weeks leading up to 29 March 2015 was 28.4%. Compared to the pie chart in Figure 5.9, what will happen to Tesco's pie slice if a chart is drawn to show market shares of British supermarkets in March 2015?

c In January 2003, for every £100 spent in British supermarkets, how many pounds were spent in Asda?

6 Table 5.3 shows the number of goods produced per week (total output) in a factory for different numbers of workers.

Table 5.3

Number of workers	Total output per week
1	0.8
2	3.6
3	6.2
4	8.6
5	10.8
6	12.6
7	14
8	15
9	15.6
10	15

a Construct a scatter diagram showing this data, with total output on the vertical axis (y-axis) and the number of workers on the horizontal axis (x-axis). Make sure the scale you choose for the y-axis allows you to plot decimals easily.

b Comment on the relationship between the number of workers and the total output per week.

6 Finance

The calculation of costs, revenue and profits is assessed at A-level only by Edexcel, OCR and WJEC/Eduqas.

CCEA assesses these topics at AS and A-level.

For AQA, the calculation of total and average costs, revenue and profits is assessed at both AS and A-level, but marginal costs, revenue and profits are assessed at A-level only.

Costs

In your economics course you will study different types of costs. One key distinction is between **fixed costs**, such as rent, which do not vary with output in the short term, and **variable costs**, such as wages, which vary with output.

The **total cost** (TC) is the total of all fixed and variable costs:

total cost (TC) = total fixed cost (TFC) + total variable cost (TVC)

The **average** total cost, average fixed cost and average variable cost are found by dividing the TC, TFC and TVC by the quantity produced, so the above equation becomes

$$\frac{\text{total cost}}{\text{quantity}} = \frac{\text{total fixed cost}}{\text{quantity}} + \frac{\text{total variable cost}}{\text{quantity}}$$

or

average cost (AC) = average fixed cost (AFC) + average variable cost (AVC)

If you are studying the full A-level in economics, you will also encounter the concept of **marginal** cost. The marginal cost is the change in total cost that results from increasing output by one extra unit. It is given by the formula

$$\text{marginal cost (MC)} = \frac{\text{change in total cost}}{\text{change in quantity}} = \frac{\Delta TC}{\Delta Q}$$

(Recall that the Greek letter Δ is shorthand for 'change in'.)

A Worked examples

a   Table 6.1

Quantity	TVC	TFC	TC	AVC	AFC	AC
0	£0	£500	£500	–	–	–
10	£200	W	X	£20	£50	Y
20	£300	£500	£800	Z	£25	£40

Copy and complete the table by calculating the values of:

i **W**

The value W is a total fixed cost. As fixed costs do not vary with quantity of output, they must be the same for the entire range of output quantities.

So W = £500.

ii **X**

The value X is a total cost. This is obtained by adding the total fixed cost and the total variable cost, so

X = TFC + TVC = £500 + £200 = £700

iii **Y**

The value Y is an average cost. This can be calculated either as AFC + AVC or as TC divided by quantity (Q). Therefore

Y = AFC + AVC = £50 + £20 = £70

or, using TC = X = £700 from part **ii**,

Y = TC ÷ Q = £700 ÷ 10 = £70

iv **Z**

The value Z is an average variable cost. This can be calculated either as TVC divided by Q or as AC minus AFC:

Z = TVC ÷ Q = £300 ÷ 20 = £15

or

Z = AC − AFC = £40 − £25 = £15

Note: worked example b is for A-level candidates only.

b **Copy and complete the MC and AC columns in Table 6.2. Give your answers to the nearest pence.**

Table 6.2

Quantity	TC	MC	AC
0	£150		–
2	£210		
4	£245		
6	£260		
8	£290		

To calculate the marginal cost column, use the formula

$$MC = \frac{\Delta TC}{\Delta Q}$$

Then the entries are:

$$\frac{(£210 - £150)}{(2 - 0)} = \frac{£60}{2} = £30$$

$$\frac{(£245 - £210)}{(4 - 2)} = \frac{£35}{2} = £17.50$$

$$\frac{(£260 - £245)}{(6 - 4)} = \frac{£15}{2} = £7.50$$

$$\frac{(£290 - £260)}{(8 - 6)} = \frac{£30}{2} = £15$$

To calculate the average cost column, use the formula

$$AC = \frac{TC}{Q}$$

Then the entries are:

$$\frac{£210}{2} = £105$$

$$\frac{£245}{4} = £61.25$$

$$\frac{£260}{6} = £43.33$$

$$\frac{£290}{8} = £36.25$$

The completed table is as follows:

Table 6.3

Quantity	TC	MC	AC
0	£150		–
		£30.00	
2	£210		£105.00
		£17.50	
4	£245		£61.25
		£7.50	
6	£260		£43.33
		£15.00	
8	£290		£36.25

B Guided questions

Copy out the workings and complete the answers on a separate piece of paper.

Note: question 1 is for A-level candidates only.

1 **The marginal cost function for a small toy manufacturer is shown in Table 6.4. Calculate the total cost values left blank in the table.**

Table 6.4

Quantity	MC	TC
0		£20
	£8	
3		
	£6	
6		
	£9	
9		

Marginal cost is the increase in total cost for every additional unit of output.

By rearranging the marginal cost formula $MC = \frac{\Delta TC}{\Delta Q}$, you can calculate the change in total cost as

$$\Delta TC = MC \times \Delta Q$$

- As the quantity increases from 0 to 3, you have $\Delta TC = £8 \times$ _____ = £ _____,
  so $TC = £20 + \Delta TC = £20 + £$_____ = £ _____
- As the quantity increases from 3 to 6, you have $\Delta TC = £$_____ $\times$ _____ = £ _____,
  so $TC = £$_____ + £ _____ = £ _____
- As the quantity increases from 6 to 9, you have $\Delta TC = £$_____ $\times$ _____ = £ _____,
  so $TC = £$_____ + £ _____ = £ _____

2 **Figure 6.1 shows the total fixed cost for a building company. The company has increased its house building from 1.5 million to 2 million this year. Calculate the average fixed cost of producing 2 million houses.**

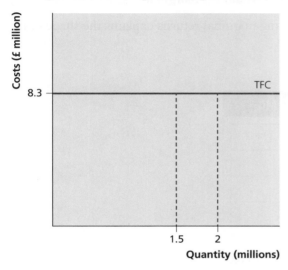

Figure 6.1

Average fixed cost (AFC) is total fixed cost (TFC) divided by quantity.

Here you are asked to calculate the AFC when the quantity is 2 million:

$AFC = TFC \div quantity =$ _____ $\div$ _____ = _____

C Practice questions

Note: question 3 is for A-level candidates only.

3 A company calculated that to increase its output from 100 to 105 would lead to a rise in total cost of £3400. Calculate the marginal cost of increase in output for this company.

4 A business is manufacturing wooden tables. The total fixed cost is £20 000 and the total variable cost to produce 100 tables is £54 000.
   a Calculate the total cost of producing 100 tables.
   b When the business manufactures 100 tables, calculate:
      i   the average fixed cost
      ii the average variable cost
      iii the average cost

5 Copy and complete Table 6.5 to show that average cost is falling over the output range 1000 to 3000.

Table 6.5

Quantity	TFC	TVC	TC	AC
1 000	£12 000	£3 000		
2 000	£12 000	£3 500		
3 000	£12 000	£3 800		

Note: question 6 is for A-level candidates only.

6 A firm is trying to find out when diminishing marginal returns* will set in. This occurs when the marginal cost begins to rise as output increases. Copy and complete Table 6.6. At the addition of which unit would the marginal cost start to rise?

(*Note: the phenomenon of diminishing marginal returns explains the shape of the typical marginal cost curve.)

Table 6.6

Quantity	TC	MC
0	£20	
1	£25	
2	£29	
3	£31	
4	£38	

7 Figure 6.2 shows a firm's average cost curve.

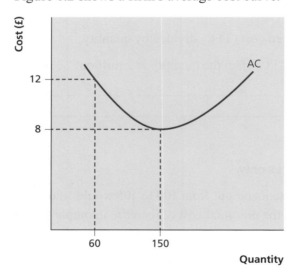

Figure 6.2

Calculate:

a the total cost at 60 and at 150 units

Note: part b is for A-level candidates only.

b the marginal cost between 60 and 150 units

Revenue

Revenue is what firms gain from selling their goods or services. In the economics course you will study different types of revenue:

- **total revenue** (TR) comprises all the revenue earned by the firm:

   TR = price per unit of product $\times$ quantity sold

- **average revenue** (AR) is the amount of revenue earned per unit of product sold:

   $$AR = \frac{\text{total revenue}}{\text{quantity}} = \text{price per unit of product}$$

If you are studying the full A-level in economics, you will also encounter **marginal revenue** (MR) — this is the extra revenue earned from selling one additional unit of output:

   $$MR = \frac{\text{change in total revenue}}{\text{change in quantity}} = \frac{\Delta TR}{\Delta Q}$$

Here are some areas of economics in which revenue concepts are used:

- **Price elasticity of demand** — this allows us to predict the relationship between price changes and total revenue. For example, lowering the price will boost revenue for goods and services with price-elastic demand (see the 'Elasticity calculations' section on page 22 in Unit 2).
- **Demand and supply** — demand and supply diagrams can be used to infer changes in total revenue. For example, they could be used to show that higher revenue is obtained when demand increases or when indirect taxes are imposed.
- **Price discrimination** — total revenue may be increased by lowering price in the elastic sub-market and raising price in the inelastic sub-market.
- **Revenue maximisation** — rather than profit maximisation, an alternative objective that a firm may have is revenue maximisation. The total revenue is maximised when marginal revenue is zero.
- **Cost and revenue diagrams** — these diagrams can be used to illustrate and compare the revenue earned in different market structures, such as perfect competition and monopoly.
- **Market share and concentration ratios** — market share is the percentage of total revenue that a firm controls, and concentration ratios give the market share of a number of top firms in a market (see the 'Percentages' section on page 15 in Unit 2).

A Worked examples

Note: worked example a is for A-level candidates only.

a   Table 6.7 shows the revenues of a small furniture manufacturer.

Table 6.7

Quantity	AR	TR	MR
100	A	£25 000	£150
200	£200	£40 000	
300	£150	B	C

i   Calculate the values of A, B and C.

- A is an average revenue and can be calculated using $AR = \dfrac{TR}{Q}$. So

$$A = \frac{£25\,000}{100} = £250$$

- B is a total revenue and can be calculated using TR = price × quantity. So

$$B = £150 \times 300 = £45\,000$$

- C is a marginal revenue and can be calculated using $MR = \dfrac{\Delta TR}{\Delta Q}$. So

$$C = \frac{(£45\,000 - £40\,000)}{(300 - 200)} = \frac{£5000}{100} = £50$$

ii   **With reference to the revenues in Table 6.7, is the firm operating under perfect competition or imperfect competition?**

Remember that, in a perfectly competitive market, as the quantity increases, average revenue remains constant and equals marginal revenue, and total revenue rises at a constant rate.

In this case, the market is imperfectly competitive, because as quantity increases, the average and marginal revenues fall, while total revenue rises at a slowing rate.

b   **A cinema used to charge all customers £10 for a ticket. It now charges different rates for adults and students, as shown by the price–demand graphs in Figure 6.3, and their overall revenue has increased since the price change.**

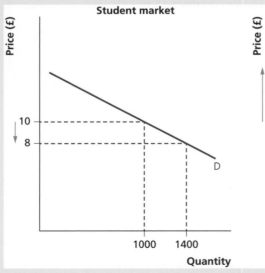
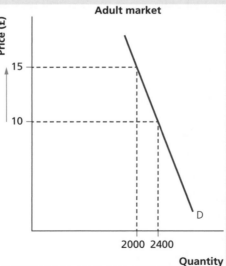

Figure 6.3

i **Calculate the total revenue of the cinema when tickets were priced at:**

   A £10 for all customers

   Total revenue is price multiplied by quantity.

   When tickets were £10 each for all customers, the left graph in Figure 6.3 shows that 1000 students bought tickets, and the right graph shows that 2400 adults bought tickets.

   Therefore the overall quantity of cinema tickets sold was 3400, so

   total revenue = £10 × 3400 = £34 000

   B £8 for students and £15 for adults

   Total revenue is price multiplied by quantity.

   The left graph in Figure 6.3 shows that when tickets were £8 each for students, the demand (i.e. quantity sold) was 1400, so

   revenue from students = £8 × 1400 = £11 200

   The right graph in Figure 6.3 shows that when tickets were £15 each for adults, the demand was 2000, so

   revenue from adults = £15 × 2000 = £30 000

   Therefore

   total revenue = £11 200 + £30 000 = £41 200

ii **With reference to price elasticity of demand, briefly explain why total revenue increased for the cinema following the price changes.**

   Total revenue has risen because the cinema has reduced price where demand is price-elastic (the student market) and increased price where demand is price-inelastic (the adult market). The cinema has used price discrimination to increase total revenue.

   To practise your skills in making elasticity calculations, you can show that the price elasticity of demand (PED) is −2 for students and −0.4 for adults. So the student market has price-elastic demand (PED below −1) and the adult market has price-inelastic demand (PED between −1 and 0). See 'Elasticity calculations' on page 22 in Unit 2 for further information.

B Guided questions

Copy out the workings and complete the answers on a separate piece of paper.

1 **The price elasticity of demand for a takeaway curry is −2.0. The shop owner decides to reduce the price of the curry by 10%, from £10 to £9 per serving. The weekly demand for the curry was 25 before the price reduction.**

   a **Calculate the original weekly total revenue.**

   Use the formula

   total revenue = price × quantity

   The original total revenue was £___ × ___ = ___

   b **Find the new weekly total revenue after the price change.**

   The question gives you the price elasticity of demand (PED) and the percentage change in price. These are two of the three values in the PED formula

   $$PED = \frac{\%\Delta \text{ quantity demanded}}{\%\Delta \text{ price}}$$

Step 1: rearrange the PED formula to make '%Δ quantity demanded' the subject.

%Δ quantity demanded = PED × %Δ price

Step 2: substitute in the given values to find %Δ quantity demanded.

Step 3: the new quantity is obtained from the rearranged percentage change formula

$$\text{new quantity} = \text{original quantity} \times \left(1 + \frac{\%\Delta \text{ quantity demanded}}{100} \right)$$

Step 4: to get the new total revenue, multiply the new price by the new quantity.

Note: part c is for A-level candidates only.

c **Find the marginal revenue following the price reduction.**

Use the formula

$$\text{MR} = \frac{\text{change in total revenue}}{\text{change in quantity}} = \frac{\Delta \text{TR}}{\Delta \text{Q}}$$

Note: question 2 is for A-level candidates only.

2 **Figure 6.4 shows the average revenue and marginal revenue for a train company. The company currently maximises profits at £50 a seat. Calculate the increase in revenue if the company decided to maximise revenue instead.**

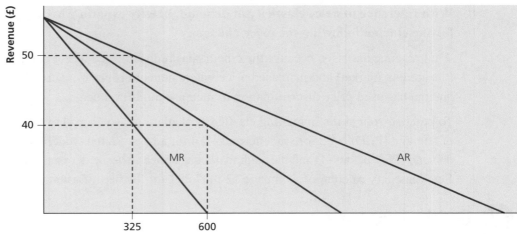

Figure 6.4

- When the firm is maximising profit, it is selling 325 seats at £50 each, so

    total revenue = £_____ × _____ = _____

- Revenue maximisation occurs when marginal revenue (MR) is zero.
  At this point:
  □ the number of seats sold is _____
     (look at where the MR graph line reaches the horizontal axis)
  □ the price of each seat is £_____
     (look at the value of the AR graph line at this quantity)
  □ so total revenue is £_____ × _____ = _____

- Find the difference between the total revenues obtained from maximising profit and from maximising revenue.

C Practice questions

3   An author has sold 25 000 copies of a new book in the first week of its release and earned £249 750. Calculate the author's average revenue.

Note: question 4 is for A-level candidates only.

4   A small commodity trader has the following revenue functions.

Table 6.8

Quantity	AR	TR	MR
5	$7.50	A	
			D
10	$7.50	B	
			E
15	$7.50	C	

a   Calculate the values of A, B, C, D and E.
b   Is the firm a price taker or a price maker? Briefly explain your answer.

5   If the government imposes a tax of 70p per unit on takeaway burgers and 1.5 million burgers were consumed during that financial year, calculate the government's total tax revenue received from takeaway burgers.

6   Figure 6.5 shows the supply and demand functions for a mobile phone app. There has been an increase in demand for the app, as illustrated by the shift from the D_1 line to the D_2 line.

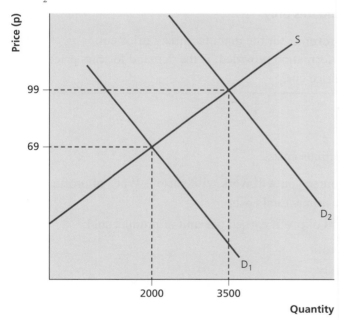

Figure 6.5

Following the shift in demand, calculate:
a   the increase in total revenue

Note: part b is for A-level candidates only.

b   the marginal revenue

7 Figure 6.6 shows the total revenue curve for a car manufacturer selling its latest model. It decides to drop the price of this car model in order to increase sales from 50 000 to 70 000.

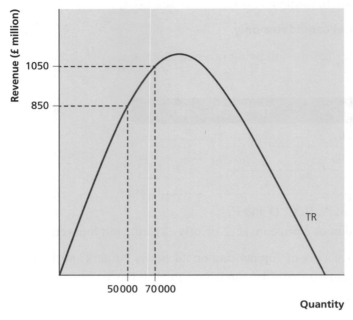

Figure 6.6

a Calculate the old price and new price of the car.

Note: part b is for A-level candidates only.

b Calculate the marginal revenue for the manufacturer's price cut.

c With reference to the information provided, is the demand for cars price-elastic or price-inelastic for this price cut?

Profit

Profit is calculated as revenue minus costs.

In the AS/A-level economics course you will work with various types of profit:

■ **total profit** is total revenue minus total costs

■ **average profit** is the amount of profit earned per unit of product sold:

$$\text{average profit} = \frac{\text{total profit}}{\text{quantity}}$$

or

$$\text{average profit} = \text{average revenue (AR)} - \text{average cost (AC)}$$

If you are studying the full A-level in economics, you will also encounter **marginal profit** — this is the extra profit earned from selling one additional unit of output:

$$\text{marginal profit} = \frac{\text{change in total profit}}{\text{change in quantity}}$$

or

$$\text{marginal profit} = \text{marginal revenue (MR)} - \text{marginal cost (MC)}$$

Here are some areas of economics in which profit concepts are used:

- **Profit maximisation** — this is one of the fundamental assumptions used in much of the **theory of the firm**. Total profit is maximised (i.e. at its highest) when marginal cost equals marginal revenue or when marginal profit is zero.
- **Perfect competition and monopoly** — supernormal and normal profits can be shown on perfect competition and monopoly diagrams.

Ⓐ Worked examples

a Table 6.9 shows the weekly revenues and costs for a small bakery. The bakery sold 1200 items during the week.

Table 6.9

Total revenue	£2 784
Total fixed costs	£750
Total variable costs	£960

i   **Calculate the weekly total profit.**

- Total profit is total revenue minus total cost.
- Total cost is found by adding together the total fixed cost and the total variable cost.

Total profit = £2784 − (£750 + £960) = £1074

ii   **Calculate the weekly average profit.**

Use the formula for average profit

$$\text{average profit} = \frac{\text{total profit}}{\text{quantity}}$$

$$= \frac{£1074}{1200}$$

$$= £0.895$$

$$= £0.90 \text{ (to the nearest pence)}$$

Note: worked example b is for A-level candidates only.

b Table 6.10 shows the revenues and profits for a small wheat farmer operating under the conditions of perfect competition.

Table 6.10

Output	TR	TC	Total profit	Marginal profit
3	£37.50	£150	−£112.50	
				£5.35
10	£125	£200	−£75	
				C
24	£300	£250	A	
				£8.33
36	£450	£300	£150	
				£0
40	£500	£350	B	
				−£12.50
42	£525	£400	£125	

i **Calculate the values of A, B and C.**

A and B are total profit values, which can be calculated as total revenue (TR) minus total cost (TC).

Therefore

A = £300 − £250 = £50
B = £500 − £350 = £150

C is a marginal profit value. Using the formula

$$\text{marginal profit} = \frac{\text{change in total profit}}{\text{change in quantity}}$$

we find that

$$C = \frac{£50 - (-£75)}{24 - 10} = \frac{£125}{14} = £8.93 \text{ (to the nearest pence)}$$

ii **With reference to the table, over what range of output is total profit maximised for the farmer?**

Profit is maximised for output ranging from 36 to 40. This is because:

- Total profit is highest (£150) at both output quantities 36 and 40.
- Marginal profit is zero for output quantity between 36 and 40.

B Guided questions

Copy out the workings and complete the answers on a separate piece of paper.

1 **Two television manufacturers have announced their annual profit results. The first sold 3700 units, with each sale making an average profit of £120. The second had a total revenue of £1 235 000 and total costs of £989 500.**

   a **Calculate the total profit of both firms.**

   - For the first firm, total profit can be calculated by multiplying the average profit by the quantity sold.

     Total profit of first firm = £_____ × _____ = £_____

   - For the second firm, total profit can be calculated as total revenue minus total cost.

     Total profit of second firm = £_____ − £_____
     = £_____

   b **Which firm made more profit and by how much?**

   Look at your answers to part **a**. Which firm had the higher total profit figure? Then find the difference between the total profit figures.

Note: question 2 is for A-level candidates only.

2 **Figure 6.7 shows cost and revenue curves for a firm, which is assumed to be profit maximising. Calculate the total revenue, total cost and total profit at the profit-maximising quantity.**

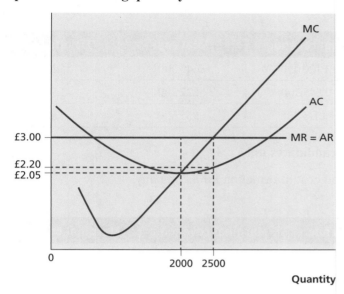

Figure 6.7

Step 1: find the profit-maximising quantity.

Total profit is maximised when marginal cost equals marginal revenue.

So, in Figure 6.7, look at where the MC curve and the MR line intersect:

MC = MR at quantity = _____

Step 2: calculate the total revenue and total cost.

From Figure 6.7, read off the average revenue (AR) and average cost (AC) at the profit-maximising quantity found in Step 1. Then multiply each by the quantity to get the total revenue and total cost.

At the profit-maximising quantity:

- AR = £_____, so TR = AR × quantity = £_____ × _____ = £_____
- AC = £_____, so TC = AC × quantity = £_____ × _____ = £_____

Step 3: calculate total profit as total revenue minus total cost.

Total profit = TR − TC = £_____ − £_____ = £_____

C Practice questions

3 During 2014, a butcher incurred total costs of £23 450 and reported a profit of £3375. How much total revenue did the butcher earn in 2014?

Note: question 4 is for A-level candidates only.

4 A manufacturer increased output from 12 000 units to 14 000 units and saw its profit decrease from £56 400 to £48 200. Calculate its marginal profit.

5 The following financial results were announced in July 2015:
- Barclays saw profit rise by 25%, up from £2.55 billion 6 months ago.
- Sky earned £1.8 billion annual profit, an 18% increase on last year.

Calculate:

a   Barclays' current profit   b   Sky's profit last year

6   Table 6.11 shows financial information for three theatres. Calculate the average profit on the tickets sold by each of these theatres.

Table 6.11

Theatre	Total revenue	Total cost	Tickets sold
The Grand	£1 120 050	£645 000	55 200
Smith & Jones	£45 300	£22 200	3 500
Theatre Company	£115 205	£55 600	7 820

Note: question 7 is for A-level candidates only.

7   Table 6.12 shows revenue and cost information for a company.

Table 6.12

Quantity	Total revenue	Marginal revenue	Marginal cost
0	£0		
10	£450		£8
20	£750		£7
30	£900		£15
40	£900		£20

a   Copy the table and complete the marginal revenue column.

b   Over what range of output is profit maximised? Briefly explain your answer.

8   Figure 6.8 shows cost and revenue curves for a local monopoly. The firm is currently profit maximising at 32 000 units of output. Calculate the firm's:

a   average profit   b   marginal profit   c   total profit

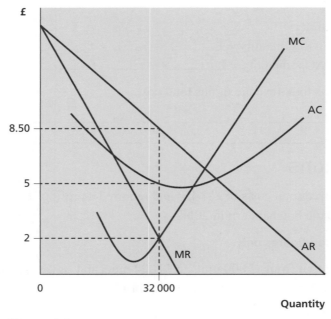

Figure 6.8

7 Written, graphical and numerical information

According to the AS/A-level economics specifications, students are expected to interpret, apply and analyse information in written, graphical, tabular and numerical forms. These skills will be needed throughout the course, across all topic areas.

This unit covers some of the general key points in interpreting and analysing information presented in various forms. The examples and questions will give you practice applying these skills to different topics in economics.

When looking at the information presented in a question, pay attention to all aspects of it to avoid any misinterpretations.

- For graphs, read the title (if given), look at how the axes have been labelled and what the units of measurement are for each variable, and note the shape of the graph — whether it is a straight line, a curve with a maximum or minimum etc.
- For tables, read the title (if given), column or row headings and units of measurement.

When writing numerical information in your answers or explanations, be sure to include all the necessary detail so that the statement is completely accurate and meaningful in context — for example, does the number need to be followed by a % sign, is the data in thousands or billions, should there be a monetary unit such as pounds or dollars?

Exam questions are likely to test whether you can interpret data accurately and draw appropriate conclusions. They may involve:

- using given information in a calculation that is based on knowledge from a specific part of the syllabus or which involves a general mathematical skill, such as working out percentage changes
- completing a table using information provided
- identifying significant features of the data
- spotting relationships between variables and giving possible explanations of why these links exist
- describing movements of data observed from graphs — often it is useful to describe the size and speed of any changes you see. Here is some useful vocabulary to get you started:
  - □ for describing size — 'considerable', 'significant', 'slight', 'negligible'
  - □ for describing speed — 'sharp', 'steady', 'rapid', 'gradual', 'slow'

- identifying trends — a trend is a general tendency or direction in the data over a specific time period. For example, an economic variable may show an upward/rising/increasing trend or, in the opposite direction, a downward/falling/decreasing trend. Sometimes a trend can be spotted quite distinctly, while in other cases there may be no obvious tendency in the data. You may be asked to provide a possible explanation of any trend you observe and comment on how it might affect other economic variables

(A) Worked examples

a   **An economy has four supermarket chains. The bar chart in Figure 7.1 shows the sales revenue for each of the supermarkets in 2000 and 2014.**

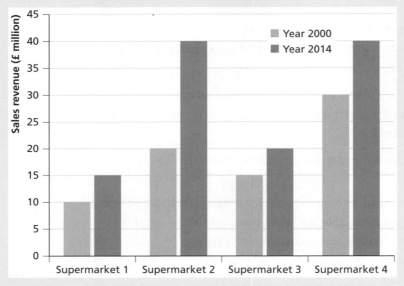

Figure 7.1 Sales revenue of the four supermarket chains

i   **What has happened to sales revenue for each of the supermarkets from 2000 to 2014?**

Sales revenue for each of the four supermarkets has risen. Supermarket 2 had the biggest proportionate rise in sales revenue (i.e. the biggest percentage change) as well as the greatest absolute rise in revenue (i.e. the increase in sales revenue in millions of pounds). Its revenue doubled from £20 million to £40 million.

ii **Work out the proportion of total sales revenue each supermarket had in 2000 (i.e. the market share).**

Step 1: add up the sales revenues of all supermarkets in 2000.

In 2000, total sales revenue = £10 million + £20 million + £15 million + £30 million = £75 million

Step 2: for each supermarket, work out what fraction of the total revenue it had in 2000.

It is best to write each fraction as a percentage, since market share is usually given in percentage terms.

Market share for supermarket 1 $= \frac{10}{75} \times 100\% = 13.3\%$ (1 d.p.)

Market share for supermarket 2 $= \frac{20}{75} \times 100\% = 26.7\%$ (1 d.p.)

Market share for supermarket 3 $= \frac{15}{75} \times 100\% = 20\%$

Market share for supermarket 4 $= \frac{30}{75} \times 100\% = 40\%$

iii **Calculate the market share for each supermarket in 2014.**

Follow the same procedure as in part **ii**, but for the year 2014.

In 2014, total sales revenue = £15 million + £40 million + £20 million + £40 million = £115 million

Market share for supermarket 1 $= \frac{15}{115} \times 100\% = 13.0\%$ (1 d.p.)

Market share for supermarket 2 $= \frac{40}{115} \times 100\% = 34.8\%$ (1 d.p.)

Market share for supermarket 3 $= \frac{20}{115} \times 100\% = 17.4\%$ (1 d.p.)

Market share for supermarket 4 $= \frac{40}{115} \times 100\% = 34.8\%$ (1 d.p.)

iv Based on the data, comment on the changes in this market between 2000 and 2014. Use your answers from parts i–iii to help draw out some key points.

The sales revenue for each of the supermarkets rose between 2000 and 2014. The most significant rise (proportionately) was for supermarket 2, whose sales revenue doubled. This resulted in a considerable increase in the market share of supermarket 2, which rose from 26.7% in 2000 to 34.8% in 2014. Supermarkets 1, 3 and 4 all experienced a slight fall in market share over this period, despite a rise in sales revenue for each. This is because total sales revenue in the market as a whole rose significantly from £75 million to £115 million, with supermarket 2 capturing a much larger slice of this total (leaving less for the others).

b In an economy, the proportion of people in employment who are women and work part-time has fallen from 20% to 15% over the past 5 years.

Explain why it may be wrong to conclude that fewer women now work part-time.

Over the past 5 years the total number of people in employment may have increased, so it could be the case that more women are working part-time but account for a lower proportion of all the people who work.

For example, if the working population 5 years ago was 10 million, then 2 million women in part-time employment would represent 20% of the total working population. If the number of people in employment has increased to 20 million now, then the number of women working part-time may have risen to 3 million, but this would only represent 15% of the current total.

This worked example highlights why, if information has been presented in percentage terms, it is important to use a word such as 'proportion', 'fraction' or 'percentage' in your written comments — otherwise they may not be accurate.

B Guided questions

Copy out the workings and complete the answers on a separate piece of paper.

1
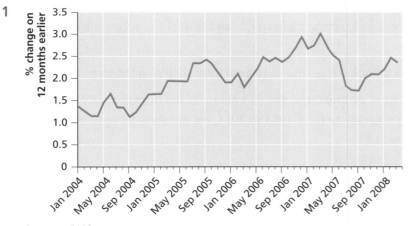

Source: ONS

Figure 7.2 Inflation in the UK, 2004–08

The following paragraph describes inflation in the UK over the period 2004–08, with reference to Figure 7.2. Decide which of the options separated by/in the brackets would describe the data most accurately and with the most suitable use of language.

This question demonstrates some of the vocabulary often used for describing graphs and the approach to take in composing written comments about economic data.

In January 2004 UK inflation was approximately 1.4%. By April 2008 it had reached approximately (2.4%/3.4%). Between January 2004 and March 2007 there was a fairly (steady/steadily) upward trend in the inflation rate, despite a few small erratic movements downwards in some months. Inflation reached a (peak/top) of (3.0%/2%) in March 2007 during this period. From March 2007 there was then a (sharp/slight) fall in inflation to approximately 1.6% in August 2007. This was the most (significant/negligible) decrease during the period, with inflation almost falling back to the January 2004 rate. After this, inflation rose (steady/steadily) again.

2 **With reference to Figure 7.3, comment on what happened to the price of oil between 2000 and 2013.**

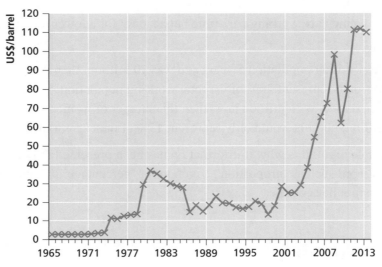

Figure 7.3 The price of oil, 1965–2013

Source: IMF

This question is about picking out main trends and features in economic data over time.

Some useful things to look for in time series data are shown in the steps below.

Step 1: get an overview by looking at the end points of the time period. Note that although the graph shows data from 1965 to 2013, the question focuses on just the period from 2000 to 2013.

What was the price of oil in 2000 compared to that in 2013?

(You can use 'approx.' when it is hard to get an accurate reading from the scale on the vertical axis.)

- Price of oil in 2000 was _____
- Price of oil in 2013 was _____

Step 2: identify the general direction of movement over the time period.

Does there seem to be an upward or a downward trend in the price of oil between 2000 and 2013?

■ Over the period there seems to be _____ trend in the price of oil.

Step 3: check for any periods during which there is more volatility in the data.

■ A time period when the price of oil was particularly volatile, with quite erratic swings, was from _____ to _____

Then describe what is happening to prices during this volatile period.

Step 4: contrast your observations in Step 3 by highlighting a period when there was a fairly consistent and steady change in the price of oil.

3   Figure 7.4 shows data relating to the UK economy.

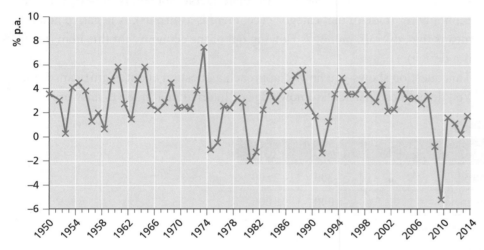

Source: ONS

Figure 7.4 Growth of real GDP, 1949–2014 (% change over previous year)

a   What is the percentage change in real GDP from 1972 to 1973?

Read off the value on the *y*-axis corresponding to the year 1973 (as this represents the percentage change over the previous year, 1972).

b   There does not appear to be a clear trend in this data. Explain why.

Look at the year-to-year movements — do they follow a general pattern in terms of their direction or are they very erratic?

c   Describe what happened to real GDP in 2010.

Step 1: read off the value on the *y*-axis corresponding to the year 2009. This gives the growth rate (percentage change) of real GDP in 2009.

Step 2: look at the sign of the growth rate. Is it positive or negative? What does this tell you about the real GDP?

Step 3: look at the size of the growth rate. This tells you by how much real GDP is changing relative to the previous year.

Step 4: based on the above information, write a sentence to explain what is happening to real GDP during 2009.

d **Explain what is likely to be happening to unemployment between 2008 and 2009.**

Step 1: look at the portion of the graph between 2008 and 2009.

What does this tell you about the change in real GDP over time during this period?

Step 2: think about how this data on change in real GDP could be related to unemployment.

4 **Table 7.1 shows the trade in services (balances) for an economy in 1999 and 2014, measured in millions of zounds (the domestic currency) using current prices.**

Identify three significant features of the data.

- Trade in services is part of the 'balance of payments'. The balance of payments records transactions between residents of one country and the rest of the world arising from international trade.
- A positive number for an item means that, overall, money is flowing into the economy for trade in that category (i.e. there is a surplus). A negative number means that, overall, money is flowing out of the economy for this item (i.e. there is a deficit).
- This question shows you how to approach a question about 'significant features'. You do not need particular knowledge of international trade to answer it.

Table 7.1

Item	1999	2014
Travel	−200	−2 500
Financial	8 000	18 000
Transportation	−3 000	−4 500
Communications	−50	25
Total	**4 750**	**11 025**

- Notice that the data is in current prices (i.e. nominal values). Think about how this will affect interpretation of the data — for instance, the values for 2014 are likely to be bigger than those for 1999, because prices have probably risen over this period due to inflation. Nevertheless, try to spot any items which have increased by a greater or lesser amount than most of the others.

Step 1: scan the different categories, paying attention to the **sizes** of the numbers and how they compare with each other.

Step 2: look at the positive and negative **signs** of all the values. What is significant about the financial item?

Step 3: look at the **proportionate** (i.e. relative or percentage) **changes** in each item from 1999 to 2014. Which item had the most significant change proportionately?

Step 4: look for any other **patterns** in the data, or any **exceptions** to general patterns and trends. For example, what is different about the communications item compared to the other items?

C Practice questions

5

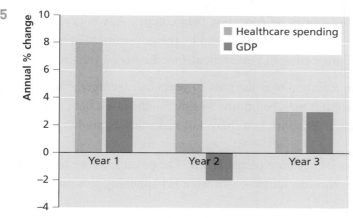

Figure 7.5 Healthcare spending and GDP

With reference to Figure 7.5, decide whether each of the following statements is true or false. (Look carefully at the axis labels and also think about what positive/negative means in this context.)

a   The percentage rise in healthcare spending in year 1 was higher than the percentage rise in GDP in year 1.

b   Healthcare spending has fallen steadily between year 1 and year 3.

c   In year 1, healthcare spending as a percentage of GDP must have risen compared to the previous year.

d   Year 2 was the only year in which GDP fell.

e   The percentage increase in total healthcare spending is falling from year to year over the period shown.

6   The pie charts in Figure 7.6 show the breakdown of government spending in 2014 for two countries, A and B. The key on the right applies to both pie charts.

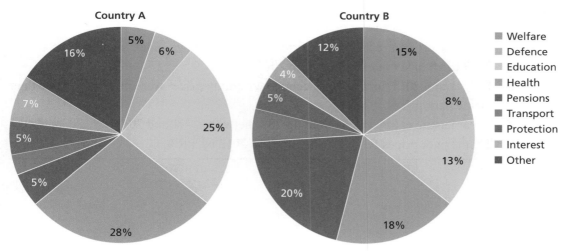

Figure 7.6

a   Which component does the government of country A spend the highest proportion on?

b   How does the proportion in part **a** compare with that for country B?

c Explain why, given the information provided, it may be wrong to state that country A spends more money on health than country B.

d If country B spends more money on education than country A, what can you conclude about the total amount of money spent by the government in country A?

e For each country, what proportion of government spending is on transport?

7 In an economy, house price inflation over the past 10 years has averaged 5%. However, house price inflation is now beginning to slow — the latest data shows annual house price inflation at 2%. This is a sharp fall and analysts predict a slowdown in the economy.

a Explain what has happened to house prices. (*Hint*: The passage above talks about house price *inflation* — this has fallen — but what is happening to actual house prices?)

b If an average house price in the economy was £125 000 this time last year, calculate the average house price now (using the given data).

8 Table 7.2 shows two price indices for a fictional economy.

Table 7.2

Year	Index of swimming prices at local leisure centres (2010 = 100)	Index of theme park prices (2010 = 100)
2009	95	98
2010	100	100
2011	110	105
2012	130	103

Decide whether each of the following statements is correct or incorrect, given the information provided in the table.

a Theme park prices went up from 98 in 2009 to 103 in 2012.

b Theme park prices went up by 5% between 2010 and 2011.

c Local leisure centre swimming prices went up from an index number of 95 in 2009 to an index number of 130 in 2012.

d In 2011 and 2012, local leisure centre swimming prices were more expensive than theme park prices.

e Local leisure centre prices went up 20% between 2011 and 2012.

f Theme park prices fell between 2011 and 2012 by 1.9%.

8 Further skills

The three topics in this unit, 'Further analysis of time series data', 'Composite indicators' and 'Seasonally adjusted figures', are assessed at AS and A-level by Edexcel only. They are not required in the specfications for AQA, OCR, WJEC/Eduqas and CCEA but would be useful knowledge for all students.

Further analysis of time series data

Note: this topic is assessed at AS and A-level by Edexcel only.

Economists collect data on many different variables, such as production levels in various industries, sizes of populations, levels of carbon dioxide emissions etc. They are interested in how these variables change over time. Such data is often presented in tables or plotted as time series graphs. We have already studied some time series graphs in Units 5 and 7. In the Worked example and Guided question 1 of this unit we will look at time series data presented in tables and demonstrate how changes over time can be expressed.

A **change** in the level of a variable can be expressed as an **absolute** change or a **percentage** (relative or proportionate) change.

Absolute changes

An absolute change is the change in the **amount**, or **value**, of the variable. Absolute changes have units, such as millions of pounds or thousands of people.

When analysing time series data, economists may discuss the absolute change, i.e. the difference in the amount of the variable, from one time period to the next.

Rates of change

From absolute changes, the **rate of change** can be calculated. This measures how much one variable changes in response to a change in another variable. In analysing time series data, the 'other' variable is time, so rate of change means **absolute change per unit time**. It is calculated by finding the absolute change in a variable over a period of time and dividing by the length of the time period. On a time series graph it is the slope, or gradient, of the line between two points on the graph. A higher value of the rate of change indicates that the level of the variable is changing more quickly over time.

It is important to be aware, however, that economists often use the word 'rate' to mean 'percentage'.

Percentage changes

The change in the level of a variable over time could also be expressed as a **percentage** change (see Unit 2). Economists often present changes in the form of percentage changes from one time period to the next, e.g. from year to year or from month to month. This shows the **relative** changes clearly and is helpful in determining whether the change over the current time period is significant compared to the level of the variable at the previous time.

Growth rates

In economics, the **growth rate** of a variable refers to the percentage change per time period (typically per year). By looking at the growth rate of a variable from year to year, economists can assess how the variable is changing over time. The speed of change can be described in terms of whether the growth rate is increasing, remaining constant or decreasing. Further words — such as 'significantly', 'sharply', 'rapidly' or 'gradually' — could be used to describe the scale of the change as compared to a different time period or another economy.

When discussing change, make sure that you use clear language in your comments. In particular, be clear whether the changes referred to in your analysis are absolute or percentage changes (growth rates) — otherwise your statements will be ambiguous.

A Worked example

Table 8.1 shows GDP data for an economy (in billions of pounds at constant prices) over a 5-year period. Comment on how real GDP is changing over this period.

Table 8.1

Year	2008		2009		2010		2011		2012
GDP (£ billion)	30		36		42		46		48
Change in GDP (£ billion)		+6		+6		+4		+2	
% change in GDP		20%		16.7%		9.5%		4.2%	

The economic variable here is real GDP.

The level of the variable is its value at a given time in units of billions of pounds. For example, the level of real GDP in 2008 is £30 billion.

Step 1: describe the **absolute changes** in the economic variable, both over the period as a whole and from each year to the next.

- Over the 5-year period, real GDP increased by £18 billion, from £30 billion in 2008 to £48 billion in 2012. There is an upward trend in real GDP.
- To visualise the year-to-year absolute changes in real GDP which make up this overall change, it is useful to plot the data on a graph:

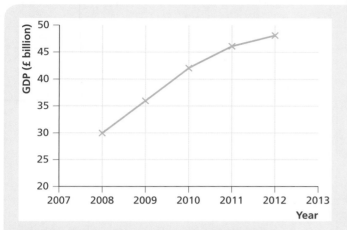

Figure 8.1 GDP at constant prices, 2008–12

Step 2: using the 'Change in GDP' row of Table 8.1 and the graph in Figure 8.1, discuss the **rates of change** in the variable.

■ Between 2008 and 2010, there is a constant rate of change in real GDP. From Table 8.1 it can be seen that the absolute change in real GDP from 2008 to 2009 and from 2009 to 2010 is the same (£6 billion in both cases). The time series graph in Figure 8.1 is a straight line between 2008 and 2010. So real GDP is increasing at a constant rate of £6 billion per year between 2008 and 2010.

■ After 2010, the rate of change in real GDP is decreasing. Table 8.1 shows that from 2010 to 2011 the rise in real GDP has fallen to £4 billion (from £6 billion previously), and from 2011 to 2012 it falls further to £2 billion. The absolute change in real GDP is still positive, so real GDP is rising, but it does so at a slower rate. In Figure 8.1, the real GDP graph gets less and less steep after 2010, indicating that the change in real GDP is slowing down.

Note that the level of a variable could also rise at an increasing rate if the absolute change from year to year is increasing over time. In this case, the time series graph would get steeper and steeper.

Step 3: describe the **percentage changes**, or **growth rates**, of the economic variable for the period as a whole and from year to year.

■ Over the period as a whole, real GDP increased by 60%: a rise from £30 billion in 2008 to £48 billion in 2012 represents a $\frac{(48 - 30)}{30} \times 100\% = 60\%$ increase.

■ The annual percentage change in real GDP was positive from each year to the next (as seen from the last row of Table 8.1) but fell consistently over the years. This means that economic growth is decreasing, i.e. slowing down, over this period: in 2009 the growth rate was 20%, but by 2012 the growth rate was only 4.2%.

B Guided question

Copy out the workings and complete the answers on a separate piece of paper.

1 Table 8.2 shows the number of part-time employees in an economy over a 10-year period. Describe the changes over this period.

Table 8.2

Year	1	2	3	4	5	6	7	8	9	10
Number of part-time employees (millions)	3	4.5	6	7.5	8.4	9	9	8.8	8.4	7
Change in the number of part-time employees from the previous year (millions)	–	+1.5						−0.2		
% change in the number of part-time employees from year to year	–	+50						−2.2		

Step 1: describe the absolute change in the economic variable over the period as a whole, and also represent this as a percentage change to give an idea of the scale of the overall change. Remember to include a negative sign for the change if the level of the variable has fallen.

Step 2: describe the changes from year to year, both the absolute changes (or rates of change) and the percentage changes (or growth rates).

To do this, first fill in the third and fourth rows of the table.

- Use the data in the third row to identify time periods where part-time employment levels have been rising or falling, and changing at a constant, increasing or decreasing rate. (Be careful how you explain the changes from year 7 to year 10.)

- Use the data in the fourth row to describe the changes, from year to year, using the growth rate in the number of part-time employees over the period shown.

C Practice question

2 Go back to Guided question 1 in Unit 5, 'Standard graphical forms' (page 54). Referring to Figure 5.6, describe how China's carbon dioxide emissions changed over the period from 1980 to 2010.

Composite indicators

Note: this topic is assessed at A-level by Edexcel only.

Many concepts used by economists — quality of life, well-being, competitiveness, social progress, development etc. — are multi-dimensional. This means that a wide-ranging set of data needs to be collected to provide a clear picture of how well an economy is performing in these areas. However, the amount and range of data can make it hard for users to get an overview quickly. Composite indicators, which combine a number of relevant individual indicators, are designed to make it easier to compare different countries and to track any changes over time within a particular country.

A composite indicator provides a quick and fairly comprehensive overview of countries' performance in a certain economic area. For evaluation purposes, however, it can be useful to consider what aspects have *not* been included in a particular composite indicator.

Composite indicators are often used by economists to rank countries. You are likely to encounter them when studying the following topics:

- **Development economics** — composite indicators such as the Human Development Index and the Social Progress Indicator help to assess a country's level of development and quality of life.
- **Environmental economics** — for example, the Happy Planet Index incorporates data on a country's ecological footprint to measure sustainable well-being.
- **Competitiveness** — for example, the Global Competitiveness Index is relevant to international trade and the performance of the economy in macro-economics.

A Worked example

The Human Development Index (HDI) is a summary measure made up of the following components:
- **health — measured in terms of life expectancy**
- **education — mean number of years in schooling, or expected years for those children entering schooling, and adult literacy rates**
- **standard of living — measured by gross national income (GNI) per head at purchasing power parity**

It treats these three dimensions equally. The highest possible theoretical value is 1 and the lowest value is zero.

Countries are ranked based on their HDI. In 2013 Norway was ranked first with an HDI of 0.944, the UK was 14th with an index of 0.892, and the Central African Republic was 185th with an index of 0.341.

i **Justify why each of the three dimensions has been included to provide this composite measurement of development for a country.**

Education is a key dimension because it signifies the capabilities that individuals have to enrich their lives and represents a form of investment in the future. The health of its people is also vital to a country's development. GNI per capita indicates the average standard of living. All of these factors are important in assessing the development of a country.

ii **Explain how a country might have a relatively high GNI per capita but a relatively low ranking on the HDI.**

A low ranking could be due to either a low score on education and/or low life expectancy in that country. For example, South Africa, despite having a relatively high income per capita, ranks lower on the HDI because it has a high prevalence of AIDS, causing life expectancy to be quite low.

B Guided question

Copy out the workings and complete the answers on a separate piece of paper.

1 **On the Human Development Index, a country with a fairly low income per head may be ranked relatively high. Explain why this can happen.**

Step 1: write down the three dimensions which are used to compile the HDI.

Step 2: given that a country has a fairly low score in one dimension, what conclusions can be drawn about the other two dimensions that make up this composite indicator if the country ends up with a fairly high ranking?

C Practice questions

2 The Human Development Index has been criticised for being too narrow since it only includes three quality-of-life indicators. Look up HDI on the United Nations website and list some factors that are not included in this index but which might affect quality-of-life and development issues.

3 The Happy Planet Index (HPI) was developed by the New Economics Foundation in 2006. It was designed to measure the extent to which a country can provide long and happy lives for its citizens in a sustainable way. The HPI ranks countries on how many long and happy lives they produce per unit of environmental input. So it considers life expectancy, how individuals in the country rate their lives and the ecological footprint of the country (broadly based on the amount of land per person needed to sustain the country's consumption patterns).

   a What are the advantages of a composite indicator such as the Happy Planet Index?

   b Why do many high-income countries score relatively low on the Happy Planet Index?

   c The lowest-income countries in sub-Saharan Africa tend to rank even lower than the countries in part **b**. What is the likely explanation of this?

   d Why would most analysts want further data to help judge the sustainable quality of life in a country?

Seasonally adjusted figures

Note: this topic is assessed at AS and A-level by Edexcel only.

Economists collect data on a variety of economic variables — such as unemployment figures, GDP, consumption, credit card lending and production volumes — and often plot the data against time to analyse how the variables change over time. See the start of this unit and Units 5 and 7 for more on time series data.

Some sets of time series data are characterised by **seasonal fluctuations** and typical effects associated with calendar dates (such as the different number of working days in a month or the timing of Easter). This part of the data is called the **seasonal component**. For data sets with a seasonal component, the data values can be expected to fluctuate with a similar intensity during certain periods of the year, if such a pattern has been observed in the past.

Economists are often interested in data from which any seasonal components have been removed, called **seasonally adjusted** data. Such data can help users spot underlying trends, and analysts will know that the causes of these movements are unrelated to seasonal variations.

For example, the unemployment rate may fall in the summer months as many industries, such as the tourism sector in the UK, take on more labour. This seasonal reduction could disguise the fact that the underlying rate of unemployment is rising.

Seasonally adjusted figures are used in many different areas of economics, such as:

- **unemployment and employment** — for example, the Office for National Statistics (ONS) publishes unemployment figures by age and duration which have been seasonally adjusted
- **supply and demand** — in some industries supply or demand will be particularly affected by seasonal factors. For example, in the UK agricultural production volumes and construction output will be significantly lower in the winter months

A Worked examples

a For each of the following variables, identify any effect that the time of year is likely to have on the data — are the data values likely to be higher or lower than in previous months?

i Retail sales in December

In December, retail sales are higher than in previous months due to the run-up to Christmas. This gives rise to a seasonal peak in retail sales during this period.

ii House construction in winter

In winter, house construction tends to be lower, leading to a seasonal trough during this period.

b For i and ii above, comment on whether the seasonally adjusted data (which shows the underlying trend) would be higher or lower than the actual data for that period.

i Actual sales data in December shows a seasonal peak. Therefore the seasonally adjusted data (i.e. underlying sales) would have lower values in December.

ii Actual construction output in winter shows a seasonal trough. This means that the seasonally adjusted data (i.e. underlying construction output) would have higher values during this period.

B Guided questions

Copy out the workings and complete the answers on a separate piece of paper.

1 For each of the following variables, explain whether the data is likely to be higher or lower than in previous months. Then state whether the data would show a seasonal peak or trough during the period identified in the question.

a The flow of credit card lending in December

Credit card lending is short-term lending and is expected to rise during periods of high consumption.

The flow of credit card lending is likely to be _____ in December than in previous months, so the data would show a seasonal _____ in December.

b **UK tourism industry employment during the summer**

Much casual labour is taken on in the tourism industry when demand is high.

Employment in the UK tourism industry is likely to be _____ in the summer than in previous months, so the data would show a seasonal _____ in summer.

c **Retail consumer spending in January**

January is the 'Christmas blues' period.

Retail consumer spending is likely to be _____ in January than in previous months, so the data would show a seasonal _____ in January.

2 **Using the same examples as in question 1, explain whether the seasonally adjusted data would be higher or lower than the actual data for that period.**

Look at your answers to question 1.

The seasonally adjusted data will take out the seasonal peaks or troughs to reflect the underlying trend.

C Practice question

3 Table 8.3 shows the volume of output (measured at constant prices in the domestic currency, which has been converted into pounds) of new private housing construction in an economy between December 2012 and June 2015. This economy has similar climate patterns to the UK. Use this data to answer the following questions.

Table 8.3

	Dec 2012	Jun 2013	Dec 2013	Jun 2014	Dec 2014	Jun 2015
Actual volume of construction output (£ million)	100	200	120	250	150	300
Seasonally adjusted volume of construction output (£ million)	140	160	168	200	210	240

a Explain why this data is likely to follow seasonal fluctuations.

b Explain why the seasonally adjusted figures for December would be higher than the actual values of new private housing construction in this month.

c Seasonally adjusted data helps users to spot underlying trends in the data more easily. What is the underlying trend in the number of new private houses built over this period?

d The seasonally adjusted output of new private housing increased from £140 million in December 2012 to £160 million in June 2013. What might explain this?

Exam-style questions

1   The price of oil increased from \$100 to £125. If oil production rose by 3%, what is the best estimate for the price elasticity of supply for oil?   **(1)**

   A   +0.12
   B   +0.25
   C   +1.2
   D   +8.33

2   A sporting retailer knows that if it reduces the price of its tennis rackets from £50 to £45, the weekly sales of tennis balls will increase from 125 tubes to 140 tubes. Based on this information, which of the following is the best estimate of the cross-price elasticity of demand (XED) for tennis rackets and tennis balls?   **(1)**

   A   +0.33
   B   −0.33
   C   −0.83
   D   −1.2

3   A car manufacturer increases production from 2000 to 3000 cars a week. As a result, the total cost increases by £750 000.

   Given the information provided, what is the car manufacturer's marginal cost?   **(1)**

   A   £75
   B   £250
   C   £375
   D   £750

4   During the recession in the USA, disposable income per head fell by 8% and sales of board games such as Monopoly increased by 6%.

   a   Which of the following is the correct YED of board games?   **(1)**
       A   −0.75
       B   +0.75
       C   +1.33
       D   −1.33

   b   Based on the information provided, explain whether board games are a normal or an inferior good.   **(3)**

5   The sterling–euro exchange rate increased from £1 : €1.20 in December 2013 to £1 : €1.28 in October 2014.

   What is the percentage increase in the sterling–euro exchange rate over this period to the nearest per cent?   **(1)**

   A   6%
   B   7%
   C   8%
   D   10%

6 Table E.1 shows the consumer price index between 2005 and 2008.

Table E.1

Year	Consumer price index (2005 = 100)
2005	100
2006	102.3
2007	104.7
2008	108.5

A worker who lost his job during the financial crisis had £2350 in his account in 2008. How much were his savings worth in real terms, to the nearest pound, using 2005 as the base period? **(1)**

A £2166

B £2245

C £2350

D £2550

7 An economic researcher constructed an index for the world oil price. The oil price was $115 in June 2014 and this was set as the base time period.

a If the index is currently 48, calculate the current world oil price. **(2)**

b In percentage terms, by how much has the oil price fallen since 2014? **(1)**

    A 48%

    B 52%

    C 55%

    D 115%

8 Table E.2 shows the inflation rate and annual percentage change in average earnings over a period of 5 years.

Table E.2

Year	Inflation rate	Annual percentage change in average earnings
1	4.9	5.4
2	5.4	6.1
3	5.0	7.3
4	5.6	6.9
5	6.3	7.7

Which one of the following statements can be inferred from the table? **(1)**

A The price level in the economy fell between years 2 and 3.

B There is an upward trend in both the inflation rate and the rate of growth of average earnings over the period shown.

C There is no evidence of cost-push inflation during this period.

D Profits for firms must have fallen between years 1 and 5.

9 Table E.3 shows the annual percentage change in GDP for a country over 4 years.

Table E.3

Year	Growth rate of GDP (%)
1	2.8
2	1.2
3	−0.8
4	−0.2

Which of the following can be concluded from the data? **(1)**

A GDP was falling throughout this period.

B GDP continued to rise in year 2, but at a slower rate of growth compared to the previous year.

C GDP in year 4 started to rise again.

D The current account must have been in surplus during years 1 and 2.

10 With an identical amount of resources, country A can produce 20 tractors or 60 lorries, whereas country B can produce 15 tractors or 30 lorries.

Which of the following statements is correct? **(1)**

A The opportunity cost for country A of producing 1 tractor is 2 lorries.

B For country B, the opportunity cost of producing 1 lorry is one-third of a tractor.

C For country A, the opportunity cost of producing 3 lorries is 1 tractor.

D Country A has the lower opportunity cost in the production of tractors.

11 Table E.4 shows tax revenue as a percentage of GDP and GDP at current US dollars for four different economies in 2012.

Table E.4

Country	Tax revenue as a percentage of GDP (%)	GDP at current US dollars ($)
Japan	10.1	5.95 trillion
Pakistan	10.1	225 billion
Thailand	16.5	366 billion
Norway	26.8	510 billion

Source: http://data.worldbank.org/indicator/GC.TAX.TOTL.GD.ZS and
http://data.worldbank.org/indicator/NY.GDP.MKTP.CD

From the data it can be inferred that, in 2012: **(1)**

A Norway had both the highest proportion of GDP collected as tax revenue and the highest GDP.

B Japan and Pakistan both collected the same amount of money in tax revenue.

C Thailand's total tax revenue was equivalent to approximately US$60 billion.

D Japan's GDP per head must be higher than that of the other countries.

12 The UK's 'big four' accountancy firms have often been under a high degree of scrutiny due to their dominance of the industry. The chart in Figure E.1 shows the market share of the top firms in the industry.

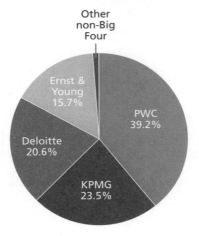

Figure E.1

Giving your answer to the nearest per cent, calculate:

a the two-firm concentration ratio **(2)**

b the four-firm concentration ratio **(2)**

13 Figure E.2 shows the breakdown of tax paid on a typical pint of beer in the UK. Giving your answer the nearest per cent, calculate the percentage of the pint made up of taxes. **(4)**

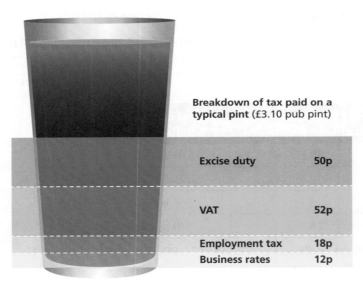

Breakdown of tax paid on a typical pint (£3.10 pub pint)

Excise duty	50p
VAT	52p
Employment tax	18p
Business rates	12p

Figure E.2

14 Table E.5 shows the consumer price index for the UK economy in recent times.

Table E.5

	Consumer price index (May 2005 = 100)
May 2013	126.1
May 2014	128.0
May 2015	128.2

Source: ONS

Giving your answer to three significant figures, calculate the UK's annual rate of inflation in:

a May 2014 **(2)**

b May 2015 **(2)**

15 Researchers have found out the following about bus and rail travel in the UK:

- Price elasticity of demand was estimated to be −0.40 for bus journeys and −0.50 for rail journeys.
- Cross-price elasticity of demand between bus and rail travel was estimated at +0.10.

Using the information provided, calculate the percentage change in the number of bus journeys if:

a bus fares were increased by 5% **(2)**

b the price of rail journeys fell by 15% **(2)**

16 Table E.6 shows the number of unemployed people and the working population in selected UK regions. Calculate how much higher the unemployment rate is in the North East compared to London. Give your answer to three significant figures. **(4)**

Table E.6

Region	Total unemployed	Working population
London	354 000	4 370 000
North East	134 000	1 300 000
North West	270 000	3 410 000
South West	187 000	2 750 000

Source: ONS

17 A self-published author has estimated the costs and revenues associated with different quantities of books sold online. The author is looking to pursue a policy of sales maximisation* to establish themselves in the market.

(*Sales maximisation is an objective where the firm tries to achieve the highest possible sales volume without making a loss.)

Table E.7

Quantity	Price	Total revenue	Profit or loss	Total cost	Marginal cost
5 000	£25			£50 000	
					7.5
7 000	£20				
					10
9 000	£15				
					12.5
11 000	£10				
					20
13 000	£5				

Copy and complete Table E.7 to find the sales-maximising quantity for the author where only normal profits are made. **(4)**

18 The graph in Figure E.3 shows the gold price since 2008. Using 13 November 2008 as the base period, calculate an index number for 5 September 2011. Give your answer to three significant figures. **(4)**

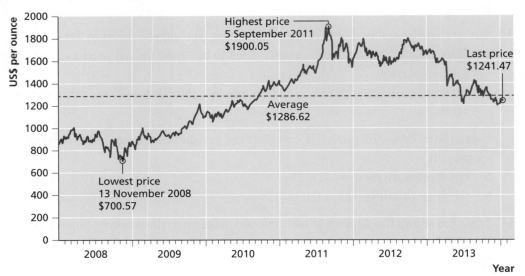

Source: Bloomberg

Figure E.3

19 A cinema estimates that it receives 10 000 visitors a week if the average price of seeing a film is £5. The cinema then reduces its ticket price to £4 and sees its total revenue increase by £14 000. Calculate the price elasticity of demand for the cinema. **(4)**

20 Table E.8 shows median weekly earnings in different regions of the UK.

Table E.8

UK regions	Median gross weekly earnings (£) of full-time employees, April 2009
United Kingdom	**489**
North East	436
North West	460
Yorkshire and the Humber	451
East Midlands	457
West Midlands	456
East of England	479
London	627
South East	514
South West	454
Wales	441
Scotland	474
Northern Ireland	439

Source: ONS Statistical Bulletin 'Annual Survey of Hours and Earnings 2009', slightly adapted.

Using the data, describe the main differences in earnings of full-time employees between the regions of the UK in 2009. **(5)**

21 Table E.9 shows the price index for food compared to the price index for all other items over a period of 4 years.

Table E.9

Year	Index of food prices	Index of 'all other items' prices
1	100	100
2	105	104
3	108	103
4	112	107

Identify two significant points of comparison between changes in the index of food prices and changes in the index of 'all other items' prices over the period shown. **(4)**

22 With reference to Figure E.4, describe what happened to the value of the pound against the Deutschmark (Deutschmarks per pound — denoted DM/£) between 1987 and 1995. **(2)**

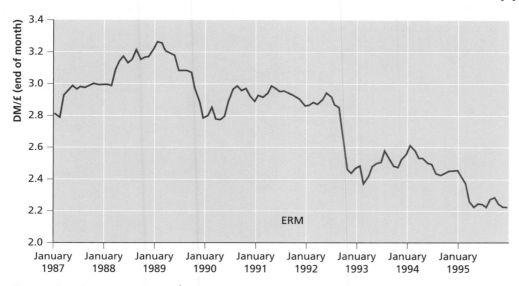

Figure E.4 The nominal DM/£ exchange rate, 1987–95

Note: the 'ERM' block on the graph refers to the UK's membership of the Exchange Rate Mechanism during this period. You do not need to know what this means to answer the question.

23 Table E.10 provides a summary of the effects of taxes and benefits on household income in the UK, organised by quintile groups, in the financial year 2013–14. Original income is income before the effects of taxation and benefits. Final income is income after the impact of taxation and benefits.

Table E.10

Average annual income per household	Bottom	2nd	3rd	4th	Top
Original income	£5 521	£13 731	£24 842	£40 880	£80 803
Final income	£15 504	£23 173	£28 358	£36 401	£60 027

Source: ONS Release 'The Effects of Taxes and Benefits on Household Income, Financial Year Ending 2014' (29 June 2015), data extracted from Figure 1.

Explain the impact on UK household income distribution, in 2013–14, of the tax and benefit system. **(4)**

(To do this, it would be helpful to calculate the ratio of average household income of the top quintile group compared to average household income of the bottom quintile group for both original income and final income. This is likely to improve the quality of your explanation.)

Appendix

Exam board cross-reference charts

With all new economics specifications, quantitative skills make up 15% of the marks at AS and 20% of the marks at A-level. These skills will be assessed across the course objectives.

In the tables in this appendix, the following symbols indicate whether or not a skill is required:

○ = **required**   ○ = **some skills required at AS**   ○ = **not required**

The information in the tables is intended as a guide only. You should refer to your specification for full details of the topics you need to know.

AQA

Unit	Topic	AS	A-level	Specification requirement
1	Fractions	○	○	'Calculate, use and understand fractions'
1	Ratios	○	○	'Calculate, use and understand ratios'
2	Percentages	○	○	'Calculate, use and understand percentages'
2	Percentage changes	○	○	'Calculate, use and understand percentage changes'
2	Elasticity	○	○	'Make elasticity calculations and interpret the results'
3	Mean and median	○	○	'Understand and use the terms median and mean'
3	Quantiles	○	○	'Understand and use the terms for relevant quantiles'
4	Index numbers	○	○	'Calculate and interpret index numbers'
4	Converting money to real terms	○	○	'Make calculations to convert from nominal to real terms'
5	Standard graphical forms	○	○	'Construct and interpret a range of standard graphical forms'
6	Costs	○	○	'Calculate total and average cost **and marginal cost at A-level**'
6	Revenue	○	○	'Calculate total and average revenue **and marginal revenue at A-level**'
6	Profit	○	○	'Calculate total and average profit **and marginal profit at A-level**'
7	Written, graphical and numerical information	○	○	'Interpret, apply and analyse information in written, graphical and numerical forms'
8	Further analysis of times series data	○	○	*Not required in specification but useful for all students*
8	Composite indicators	○	○	*Not required in specification but useful for all students*
8	Seasonally adjusted figures	○	○	*Not required in specification but useful for all students*

Edexcel

Unit	Topic	AS	A-level	Specification requirement
1	Fractions	○	○	'Calculate, use and understand fractions'
1	Ratios	○	○	'Calculate, use and understand ratios'
2	Percentages	○	○	'Calculate, use and understand percentages'
2	Percentage changes	○	○	'Calculate, use and understand percentage changes'
2	Elasticity	○	○	'Make elasticity calculations and interpret the results'
3	Mean and median	○	○	'Understand and use the terms median and mean'
3	Quantiles	○	○	'Understand and use the terms for relevant quantiles'
4	Index numbers	○	○	'Calculate and interpret index numbers'
4	Converting money to real terms	○	○	'Make calculations to convert from nominal to real terms'
5	Standard graphical forms	○	○	'Construct and interpret a range of standard graphical forms'
6	Costs	○	○	'Calculate cost (total, average, marginal)'
6	Revenue	○	○	'Calculate revenue (total, average, marginal)'
6	Profit	○	○	'Calculate profit (total, average, marginal)'
7	Written, graphical and numerical information	○	○	'Interpret, apply and analyse information in written, graphical and numerical forms'
8	Further analysis of times series data	○	○	'Distinguish between changes in the level of a variable and the rate of change'
8	Composite indicators	○	○	'Understand composite indicators'
8	Seasonally adjusted figures	○	○	'Understand the meaning of seasonally adjusted figures'

OCR

Unit	Topic	AS	A-level	Specification requirement
1	Fractions	○	○	'Calculate, use and understand fractions'
1	Ratios	○	○	'Calculate, use and understand ratios'
2	Percentages	○	○	'Calculate, use and understand percentages'
2	Percentage changes	○	○	'Calculate, use and understand percentage changes'
2	Elasticity	○	○	'Make elasticity calculations and interpret the results'
3	Mean and median	○	○	'Understand and use the terms median and mean'
3	Quantiles	○	○	'Understand and use the terms for relevant quantiles'
4	Index numbers	○	○	'Interpret index numbers **and calculate index numbers at A-level**'
4	Converting money to real terms	○	○	'Make calculations to convert from nominal to real terms'
5	Standard graphical forms	○	○	'Construct and interpret a range of standard graphical forms'
6	Costs	○	○	'Calculate cost (total, average, marginal)'
6	Revenue	○	○	'Calculate revenue (total, average, marginal)'
6	Profit	○	○	'Calculate profit (total, average, marginal)'
7	Written, graphical and numerical information	○	○	'Interpret, apply and analyse information in written, graphical and numerical forms'
8	Further analysis of times series data	○	○	*Not required in specification but useful for all students*
8	Composite indicators	○	○	*Not required in specification but useful for all students*
8	Seasonally adjusted figures	○	○	*Not required in specification but useful for all students*

WJEC/Eduqas

Unit	Topic	AS	A-level	Specification requirement
1	Fractions	○	○	'Calculate, use and understand fractions'
1	Ratios	○	○	'Calculate, use and understand ratios'
2	Percentages	○	○	'Calculate, use and understand percentages'
2	Percentage changes	○	○	'Calculate, use and understand percentage changes'
2	Elasticity	○	○	'Make elasticity calculations and interpret the results'
3	Mean and median	○	○	'Understand and use the terms median and mean'
3	Quantiles	○	○	'Understand and use the terms for relevant quantiles'
4	Index numbers	○	○	'Calculate and interpret index numbers'
4	Converting money to real terms	○	○	'Make calculations to convert from nominal to real terms'
5	Standard graphical forms	○	○	'Construct and interpret a range of standard graphical forms'
6	Costs	○	○	'Calculate cost (total, average, marginal)'
6	Revenue	○	○	'Calculate revenue (total, average, marginal)'
6	Profit	○	○	'Calculate profit (total, average, marginal)'
7	Written, graphical and numerical information	○	○	'Interpret, apply and analyse information in written, graphical and numerical forms'
8	Further analysis of times series data	○	○	*Not required in specification but useful for all students*
8	Composite indicators	○	○	*Not required in specification but useful for all students*
8	Seasonally adjusted figures	○	○	*Not required in specification but useful for all students*

CCEA

Unit	Topic	AS	A-level	Specification requirement
1	Fractions	O	O	'Understand, use and calculate fractions'
1	Ratios	O	O	'Understand, use and calculate ratios'
2	Percentages	O	O	'Understand, use and calculate percentages'
2	Percentage changes	O	O	'Understand, use and calculate percentage point changes'
2	Elasticity	O	O	'Calculate elasticity and interpret results'
3	Mean and median	O	O	'Understand and use the terms median and mean'
3	Quantiles	O	O	Understand and use relevant quantiles'
4	Index numbers	O	O	'Calculate and interpret index numbers'
4	Converting money to real terms	O	O	'Convert from money to real terms'
5	Standard graphical forms	O	O	'Construct and interpret a range of standard graphical forms'
6	Costs	O	O	'Calculate cost (total, average, marginal)'
6	Revenue	O	O	'Calculate revenue (total, average, marginal)'
6	Profit	O	O	'Calculate profit (total, average, marginal)'
7	Written, graphical and numerical information	O	O	Interpret, apply and analyse information in written, graphical and numerical forms'
8	Further analysis of times series data	O	O	*Not required in specification but useful for all students*
8	Composite indicators	O	O	*Not required in specification but useful for all students*
8	Seasonally adjusted figures	O	O	*Not required in specification but useful for all students*